Expanding Earth, Constant Mass

Expanding Earth, Constant Mass

A Revolutionary Theory of the Origins of Our
Planet and the Nature of the Universe

David W. Knight

Copyright © 2018 by David W. Knight

All rights reserved, including the right to reproduce this book or portions thereof in any form whatsoever. For information write to the publisher at the address below.

Meadow Point Press
20 Springwood Drive
Lawrenceville, NJ 08648
www.meadowpointpress.com
Printed in the United States of America

Publisher's Catalog-in-Publication data
Knight, David.
The Expanding Earth / David Knight.
    p. cm.
ISBN 978-1-387-87038-7
1. General science. 2. Geoscience. 3. Cosmology. II. Title.

First Edition

Contents

Acknowledgments

Table: Eons of Geologic Time

Chapter 1	The Earth and the Universe	1
Chapter 2	There Are Many Electrets in Space	13
Chapter 3	The Development of a Young Planet	33
	Processes That Took Place Within the New Earth	*38*
	Changes to the Inside Affect the Outside	*41*
Chapter 4	Electrets in the Cosmos	54
	Light and the Red Shift	*54*
	Supernovas	*61*
	Comets	*64*
Chapter 5	Evidence Today of the Earth's Expansion	67
	Expansion Created Geological Records of All the Periods of the Paleozoic	*67*
	The Original Continents Comprised Several of the Continents We Know Today	*71*
	Hess, Ewing, and the Age of the Oceans	*73*
	Volcanoes and the Andesite Line	*77*

Chapter 6	Other Mysteries Explained by the Earth's Expansion	**80**
	The Earth's Magnetic Field	*80*
	Gravity Was Greater When the Earth Was Smaller	*85*
Chapter 7	Ionite	**89**
Chapter 8	Challenging the Scientific Consensus	**95**
	The Great Scientists	*95*
	Rocking the Boat	*97*
	Ideas of Mine That Run Counter to Accepted Theory	*104*
Chapter 9	Final Thoughts	**109**
Appendix	The Math	**121**

Acknowledgments

I first want to thank my wife, Marian, who gave me the space to pursue my decades-long inquiry into the expanding earth.

I am deeply indebted to Archimandrite Juvenal Repass, whom I knew as John before he joined the priesthood of the Orthodox Church in America. He provided the mathematical calculations that proved that a highly charged electret could be a stable form of matter. Without that breakthrough my quest to understand how the earth expanded would have stalled a long time ago.

I am very grateful to my son Paul and his wife's brother-in-law, Tom Bontrager, for their help in editing this book and preparing it for publication.

I also thank my granddaughter, Emily Fedorko, for designing the book's cover.

And finally, I deeply appreciate my younger son, Richard, my daughter, Elizabeth, Kit and Fritz Ober, and all my other friends and relatives who generously listened to me hold forth on the topic of the expanding earth over the past many years. I am eternally grateful to them for being my sounding board.

EONS OF GEOLOGIC TIME

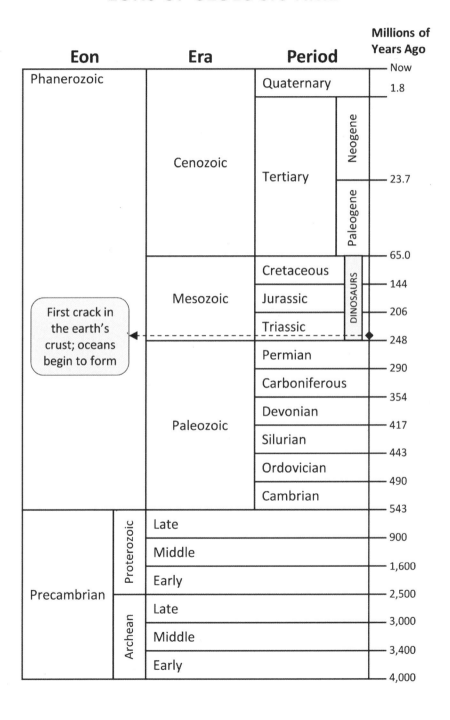

Chapter 1 The Earth and the Universe

I am very lucky. My wife has told me I shouldn't say that, but still I think I'm very lucky. We live in a very good neighborhood in a small town in a country with a strong economy and a good government. I am retired from a company that paid me so well that I was able to raise and educate a family, and enough so that my wife could put food on the table. All my children went to college. Two are happy in their marriages, and one is living alone. I have three grandchildren, all employed and happy. My wife is well, and I am writing this little book.

I used to do engineering for a small company, and I liked my work. The firm made machinery for producing small metal parts, though this book is not about machinery or small metal parts. I was a development engineer for that small company and worked much of the time alone.

My wife used to say the company would go out of business if I left them. They *have* gone out of business, but not because I left; the real reason is that few small metal parts are now made in our country.

Some years ago, I made a scientific discovery. It wasn't in my line of work; rather, it was about the earth and the universe. Part of my discovery was that the earth started out as a much smaller globe of perhaps half its present size, and expanded to its current dimensions. I was not alone in reaching that conclusion, although it is not the prevailing view. But the part of my discovery that would surprise almost everyone is that *the smaller version of our planet already had nearly as great a mass as it has now.* In this regard, I was—and remain—an outsider. A few people have come over to my side of the expanding-earth debate, but most continue to think the earth was always the same size it is now.

Chances are you believe that, too. I hope to change your mind.

I understand that Newton said something to the effect that he was "standing on the shoulders of giants" who had gone before him. I too have relied on the work of many of our scientists to develop my ideas. Scientists generally may be reluctant to believe the earth expanded, however, because so many of their hard-won ideas would have to change. From the point of view of the theories they have worked out, the earth could not have expanded. Nevertheless, I believe it did. In pursuing this idea, I found out a great deal more about our universe that goes far beyond the expansion of the earth. So much so, in fact, that finding a good title for this book has been difficult. If I had been a professional scientist and found that the rest of my community rejected my thinking, I might have abandoned my research, but that never happened. And once I was sure I was on the right track, there was no turning back.

As an undergraduate at Hobart College in upstate New York, I studied math, chemistry, and physics as a way of getting into science. At the time, I had little interest in the finer things in life, but Hobart is a liberal arts school, and I did get a well-rounded education in the end. It wasn't until graduate school at Rensselaer Polytechnic Institute (RPI) that I studied physical geology.

It was another ten years after grad school before I discovered the expanding earth for myself. About a year after that, I learned of a geologist in Hobart, Tasmania, Samuel Warren Carey, who also thought the earth had expanded. I corresponded with Professor Carey briefly before he died. He wrote several books on the subject but never correctly determined the cause of the expansion, which I also didn't know at the time. He made the mistake of trying to guess about it, and the scientific community came down on him. I later determined that he was right about the expanding earth but mistaken about the cause. Nonetheless, Carey was one of the early proponents of the expansion hypothesis. Now quite a few of us agree with him, though we're still a small minority.

From early on, I assumed that the earth's mass was nearly what it is now. If you do a search on Google, you'll find that many accept the expanding earth, yet they tend to focus their attention only on the earth's surface. Instead, I became fascinated with what could have happened within the earth to cause the expansion. The earth as it now exists is only a small part of a much larger problem, as you will see. If the earth indeed expanded from a smaller planet to its present size, then I had to determine where all its mass had come from, since I knew it could not have gained a great deal of additional mass once it had formed. This was a terrible problem for science and still is. It took me many years to find the answer, and I made my share of mistakes along the way. How the earth formed with its initial great density became an obsession—or a hobby, at least—and in the course of my explorations I did confirm that the earth expanded while maintaining nearly constant mass.

This book, then, is the story of my quest to learn how a ball of matter much smaller than the earth we know today but with all its present mass could have formed in the first place, and how it expanded to become the living earth we know today. In making sense of this, I had to develop a new theory of the formation of the solar system, for the simple reason that the accepted theory doesn't yield a small, extremely dense planet. I can say this with greater confidence now, having worked out many of the details, yet even at the beginning of my journey I had the distinct advantage of knowing that the solar system had to have formed in such a way as to produce a smaller, denser earth. What a boon it is to know a crucial end result when starting your work!

There have been many attempts to work out how the solar system formed. None has it right, and in my own attempts, too, there have been failures. My fundamental conviction about expansion led to a surprising series of conclusions, both within our solar system and without. It was a great thing to find out how the sun formed, for example, yet that would turn out to be only one of the first of my extraterrestrial discoveries. The story told by this book, you see, will be about our universe as well as our sun and earth.

I found, for instance, what caused the so-called "red shift" of distant celestial light sources. The red shift, you may remember, is the dramatic increase in the wavelength of light coming from distant objects, in which their apparent color is shifted toward the red end of the spectrum and in which the farther away those objects are, the greater the shift. The accepted explanation is that this phenomenon is the result of distant objects moving away from us at great speeds. I now know that this not what causes the red shift.

Other far-reaching discoveries emerged, not the least of which was the truth about the Big Bang. The English astronomer Fred Hoyle was never able to prove that there was no Big Bang, although he was convinced there was none. Now I know he was right. I believe I'm the first to know an array of important scientific truths: there was no Big Bang, time has been going on forever, and the universe goes on beyond what we supposed were its dimensions—indeed, it has no limit in size. These conclusions and more follow from an understanding of how the earth expanded.

I should mention how I came to realize that the earth had expanded. When my eldest son, Paul, was in grade school, he was given one of those school flyers on science. I read it to find out what they were teaching him, and there I learned that Harry Hess (a geologist at Princeton) had put forth a theory that tried to explain why the ocean floor was spreading in the center of the ocean, along a line we now know as the mid-ocean ridge. He thought the floor of the ocean was rising at the center and going down under the continents at the coasts, in a conveyor-belt sort of motion. He called it "seafloor spreading," and his account gradually developed into the theory of plate tectonics that is popular now.

I believed Hess could not have been right in his thinking. I had little to go on, but I thought it was entirely plausible that the ocean floor had simply pulled apart. At the time, I was not assuming that the earth would have to expand in order for this to happen, but I also was not assuming the impossibility of such a thing. Since there was no one to interrupt my musings, I could think as I pleased. (My son was in the fifth grade at the time. He has recently retired, so I've been at this a long time.)

Keep in mind that, at first, I had only been interested in what my child was being taught in school. But when I came to suspect that the earth must have expanded, of course I realized that scientifically it would be a big deal if I was right. I couldn't put the thought aside, and the next evening I checked out all the books on geology from our local library. In the course of my readings, I found that we did have the tools to show that the earth had expanded but that the problem of the earth's mass had convinced the scientific community that this could not have happened.

While reading *Physics Today* magazine in that same library, I found an article about the geology professor I mentioned earlier, S. Warren Carey of Tasmania, and his work on the subject of the expanding earth. I looked him up and learned a great deal. What convinced me he was on the right track was a visual experiment he conducted, in which he used a large bowl to represent a portion of the surface of the earth and placed cut-outs he had made in the bowl to represent the continents. The edges of the continents were to scale and true, and Carey left space to represent the oceans as they now exist. He found that the cut-outs would not fit together precisely to form a larger continent, but that if he made some accommodation by bending their edges and vertices into a smaller curvature, such as they would have had on the surface of a smaller earth, then they did fit. He then realized he'd made a stunning discovery: the earth must have expanded. He organized a symposium, eager to announce his finding, but his fellow geologists rejected it. Too much established science, they thought, would have to change on the assumption that the earth expanded. (But it did.)

Carey deserves a great deal of credit for his work. He did everything he could to convince the scientific world that the Earth had expanded, even though they wouldn't listen. He wrote a number of books on the subject and had others working with him. It turns out he was not the first proponent of expansion, just the one whose insights convinced me he was correct. It's true that much common wisdom would have to change in order for the scientific establishment to accept that the earth expanded, and more yet will have to change in order for them to accept my ideas, too. It may happen long after I have died, but eventually I believe that scientists will come to accept the earth's expansion and everything that flows from that fact.

John Repass, the son of a couple who were friends of ours, helped me with the math to support my theory. I'll come back to John and his invaluable contribution later.

Another fellow, John Sebas, had good reason to take me seriously and listen to what I had to say, being at the time my daughter's boyfriend. We had long conversations, and of course I liked him. My regret is that I almost kept him from being a doctor. He made the mistake of taking a college course in astronomy and writing a paper based on my ideas. His professor was a good scientist who could not accept what he'd written. Ouch. As a result, John didn't get his usual 'A' in the course. The good news is that he was admitted to medical school in spite of that grade and is now an M.D.

I want to give you some reason for continuing to read this book, but there's no one thing I can say that will persuade you. Ultimately, the only thing that will convince you this book is worth reading is the progression of facts and reasons that I will present in the chapters to come. Years ago, when I got my first computer, my son-in-law asked me why I hadn't written a book about my discoveries. I told him it was because there were some one hundred points to be made, but not to worry, I was working on it. He asked how long it would take to write each item. I told him about two weeks, more or less. He said, "OK, then in about two years you should finish your book, and I'll read it when you're done." The deadline helped, and I finished the first draft in nine months. As promised, he read it. I knew nothing about writing. I hope I've improved since then, but at the time reading the whole book was a good deed on my son-in-law's part. What you have before you is a shorter version of that first effort.

Now I should say a bit about myself. I've been working on these geophysical and astrophysical topics for a long time. One reason I've taken so long to publish, apart from a daunting sequence of facts and reasons that have had to be carefully assembled, is that I know that few in the scientific community will even entertain my ideas. For example, in order for my account of the formation of the oceans and continents to be correct, many of our most respected seismologists would have to have things completely backward. I believe I now understand how the accepted theory got turned around. You may know that it's impossible to tell from a seismographic reading the direction in which the first motion of a seismic disturbance took place. When it came to the direction of movements of the sea floor at its juncture with the continental shelf, no one in the scientific community would consider any possibility but the one they already believed to be the case—namely, that the sea floor was moving downward, under the continental shelf. To suggest that it was moving upward, out from under the shelf, would have flown in the face of the well-established consensus. No seismologist was going to entertain an alternative theory that everyone "knew" had to be false.

However, if you first accept that the earth is expanding, then it becomes perfectly understandable that the continent is simply pulling away from the mantle because of the earth's expansion, leaving the ocean floor to rise to its natural level. As I'll argue in more detail later, *the ocean floor is moving up from beneath the continents, not down under them.* It's so very easy to say this, but first one has to know about the expanding earth.

I dislike having to make anyone wrong, yet in the coming chapters I'll be making many good scientists wrong. My understanding of the reasons for doing so doesn't relieve my discomfort. Some will regard my criticisms of received theory as a slight no matter what.

A bit more about me. I studied physics and thought I did quite well. No, I wasn't the best. I'm afraid I wasn't as good a student as I should have been. Like many others, I wish I'd studied harder and learned more, but I'm very glad I had the opportunity to learn what I did. Hobart College is a liberal arts school, and I ended up taking all the sciences they offered. Later, in graduate school at RPI, my focus became physics, although it had been my chemistry professor who helped me get accepted.

I love the sciences but recognize that I'm not a professional scientist. I wish I were. I respect the hard work required. I have, however, used the work of professional scientists to accomplish everything I have. I've done almost no experiments myself, although I've based my conclusions on hard facts wherever possible, and I've tried to be scrupulous in separating fact from assumption. All my results can be traced back to my knowledge of the expanding earth. As will be clear to you by now, I think I'm right about that. But who am I? I'd love to sell myself as some wonderful oracle or thinking machine, but I'm just me: a good mechanic with a degree in physical sciences and some geology under his belt. That's hardly enough for omniscience. I wish I could be everything you would expect of someone who managed to answer the great questions of the universe. Failing that, I must ask you to do me the good deed of continuing to read this book with an open mind.

I should also confess that my work has been prompted by certain suspicions and predispositions. For a long time, I've found it impossible to accept the idea of a Big Bang. I never liked the standard theory of solar-system formation or the notion that the subcontinent of India once traipsed across the ocean to push up the Himalayan mountains, and I've always been suspicious of the assumption that the red shift implies an expanding universe. If I've been able to determine the truth about these phenomena, it may just be dumb luck.

Well, as I said at the outset, I am very lucky.

Chapter 2 There Are Many Electrets in Space

An electret in deep space is an unknown quantity even to the most sophisticated astronomer. There are many kinds of electret. The sort I will be discussing is a *non-conducting dielectric material* that is charged positively internally, where *the displaced electrons are on its surface*. There are no electrets of this kind on earth, and this may be why they are not widely recognized. These deep-space electrets are produced through the normal action of cosmic rays on normally occurring non-conductors such as stone dust.

Now I must ask you to suspend any disbelief you may experience until I've had a chance to explain how the relevant materials and behaviors fit together into a coherent process. Each step should follow logically, so that the progression as a whole will be easy to follow. I was forced to reject much of the theory developed over the years by others and find an alternative way of explaining the hard-won facts accumulated by our scientists. I realize it seems bold of me to say I can improve on the standard theories, but bear with me. I'm afraid that reading this book may be difficult at first, because what it has to say will be new to almost everyone. I'd like nothing more than to just open your skull and pour all I've learned into your brain. Since that's not possible, you're going to have to read and absorb each idea as we go along. Great science writers have more than just a grasp of the subject matter, they have the ability to put their thoughts on paper in such a way that the reader understands and remembers what the scientist has come to know. I don't pretend to be a great science writer, but I'll do my best to lay out my ideas in an orderly fashion, giving my reasons for arriving at each idea along the way.

The earth expanded, and this could not have happened according to all we thought we knew. I won't be able to explain the formation of the earth immediately. First, I'll need to describe what happens in space to allow a planet to form as a small, dense body. This will require outlining a *new process for the development of our sun or any star*, for the simple reason that star formation under the current theory could not have led to a small, dense earth with nearly the same mass that it has today.

In deep space, there are many dust particles, most of them just tiny pieces of rock. You are likely aware of what is commonly called "cosmic dust" or "space dust." These are small, perhaps a tenth of a millimeter in diameter. In

There Are Many Electrets in Space

space there are also many cosmic rays flying at high velocity in all directions. It is well known that a cosmic ray is really the nucleus of an atom—specifically, the nucleus of a hydrogen atom, which consists of a single proton—flying so fast that the atom's electrons have been stripped away. Some cosmic rays enter our atmosphere, but I am most interested in those far from our earth and sun, in the extreme cold of deep space.

Since in deep space there are many dust grains as well as many cosmic rays flying about in all directions, there is a likelihood that in time a cosmic ray will strike a dust grain, and such collisions will take place many billions of times. When a collision occurs, one of three things will happen: (1) the cosmic ray particle will bounce off, (2) it will pass through the grain, or (3) it will enter the grain and remain there. If it lodges inside (3), it will charge the grain positively. (A cosmic ray always carries a positive charge, since it is an atom that has lost its negatively charged electrons.) If the grain is a non-conductor, which most are, being merely tiny pieces of rock, then after many cosmic-ray strikes in which these nuclei become embedded in the grain, its positive charge will naturally build. In time, the charge will get very large. On the earth, such a charge would escape, but in deep space it will not, simply because there is no matter nearby for the charge to jump to.

Some time ago, with the help of John Repass, I found that no matter how great the positive charge in the center of the grain became, the grain would not explode.

It required solving a simple differential equation that John set up, based on information I had given him. What happens is that the electrons on the outside press inward with a force greater than the pressure pushing out. This condition of net inward pressure is of little importance when there is only a single, small grain, but it becomes more and more significant as grains combine to become much larger bodies. *When a dust grain is charged in this way, it becomes a positively charged electret.* The inward-acting pressure from the halo of electrons increases in proportion to the size of the positive charge within the grain. Picture, if you will, a small non-conducting grain of dust charged positively on the inside and surrounded by electrons that have collected on the outside. What happens over a very long time, as I will explain, is that these grains combine with one another to form astonishingly large electrets that remain charged positively on the inside yet are kept intact by electrons pressing in from the outside. Electrets of this sort can become as large as a planet. These gargantuan electrets are subject to unimaginable pressures, great enough to keep even a planet compressed into a small volume. That hypothetical piece of information is really incidental to my story, but it fits with my real discovery, that our originally much smaller earth started its life as a compact electret held to a small size by electrons pressing in from the outside. What remains is to tell you how a great many grains must have combined to form this compact planet. I will eventually explain how an electret with the mass of the earth was able to grow and change into the vibrant, living planet we know today. This in itself is a long and interesting story, but first I must tell you what I think happens to cause a miniscule charged electret to combine with similar particles to become a large, compressed body.

In the extreme cold of deep space, through cosmic-ray strikes, the charge on a lone dust grain will grow to immense proportions. And while the enlarged grain is irregular in shape, the electrons on the outside form a perfect sphere. (This is conjecture on my part, but I do believe it happens.) In physics, a perfectly spherical arrangement of matter is nearly impossible, but the halo of an electret is an exception.

Picture, if you will, a perfect sphere of electrons surrounding an electret grain. One would expect that, when another cosmic ray embeds its positive nucleus in the grain, an additional electron would be gathered from space into the electron sphere surrounding the grain, but in some cases, the existing sphere cannot accommodate an additional electron. There is a law in quantum mechanics, the Pauli exclusion principle, that determines the number of electrons that can occupy an electron shell. (An electron shell consists of all the electrons in an atom that share the same principal quantum number, or energy level. In non-quantum terms, electrons in any given shell may be thought of as being the same distance from the center of charge.) The exclusion principle states that there can be no more than a certain number of electrons in any given shell, with the maximum number varying from shell to shell. I believe this exclusion principle applies to the electrons surrounding an electret grain as it does to atoms. There is a limit to the number of electrons that can reside in the spherical electron shell at the surface of an electret in space. When the outer shell is full and another cosmic ray strike occurs, an additional electron is recruited from space to balance the charge. But that electron cannot embed itself within the existing sphere; it has to sit further away from the surface of the sphere in a new outer shell. ("Sit" is only a metaphor for describing the behavior of electrons, since we know they are always moving, but their movement makes little difference here, because we are interested in their average distance from the center.) If this creation of a new shell can happen once, it can happen again and again. Billions of strikes later, a giant halo of electrons has formed around the grain. With each new shell, the electrons in the outermost shell are held with less and less attractive force, and they define a larger sphere, so that the vertical distance between the electrons increases. Yet every electron is still attracted to the central charge within the dust grain. In short, the halo of electrons thins out as it gets larger. This halo has some serious consequences for the future of that grain and for our understanding of the universe. Among these is the fact that the electron halo refracts light passing through it to an infinitesimal degree, causing the light to lose a tiny amount of energy. I'll describe the profound implications of that phenomenon in Chapter 4.

If electrons form a halo around one dust grain, they will do the same for others—perhaps for all dust grains, given enough time. Let us say that two of these grains surrounded by electron halos approach each other. The surfaces of the enclosed grains cannot touch directly, because their electron halos are negatively charged and will repel each other. They can only approach until their halos almost touch. Let us imagine what will happen in deep space, where there are many dust grains hovering and moving about. Through our great telescopes, we can see that many grains of cosmic dust do not clump together but rather form clouds at various locations within distant galaxies.

The electret grain I am proposing consists of a dust particle with a positively charged core surrounded by electrons, so that the whole electret is like a neutral atom. I see no reason why Van der Waals forces would not act between electret grains of this sort. (It's a novel statement but likely to be true, since there is no reason it should not be.) The grains I've described are very like neutral gas atoms. A famous discovery that neutral gases do not solidify at the very low temperatures our "laws" predicted led to the discovery of Van der Waals forces in the first place.

Let us imagine a cloud of these grains, each surrounded by its halo of electrons. Let us say a stray electret grain flying through space encounters this cloud. That particle will fly into the cloud and be trapped. This will happen, because the electrons surrounding the invading grain will interfere with the electrons of the cloud particles to create a kind of viscosity. As a result, the cloud will increase slightly in size, and this kind of entrapment will happen many times over. The growth of the cloud can eventually lead to a cloud of electret grains with the total mass of our sun—indeed, that *becomes* our sun, our planets, and the rest of our solar system. But I am getting ahead of myself.

Van der Waals forces are not like gravity, in that gravity drops off with the inverse square of the distance between the acting bodies, whereas Van der Waals forces drop off much more rapidly (though they never go to zero). At even a moderately small distance, there is almost no Van der Waals attraction between two bodies, although at extremely short distances Van der Waals forces become quite effective. The reason a gecko can walk on the ceiling is due to Van der Waals forces acting on tiny hairs on its strangely shaped toes, when they are in contact with the ceiling.

Van der Waals forces act within an enormous cloud of electret particles in a very strange way. I will use a common example to illustrate what I think happens. In a rubber balloon, each rubber molecule pulls on the molecule next to it. The inward-acting pressure keeping air inside the balloon is due only to this sideways force acting between molecules. The analogous inward force created by Van der Waals forces pulling together dust grains on the periphery of a dust cloud is tiny, but it is real. In the cloud, we have dust grains at a certain distance from the center attracting one another laterally; the attraction among dust grains above and below does not count regarding this lateral, containing force at a given level.

Now imagine a spherical layer of particles within this very large cloud, with its center at the cloud's center—a "balloon" surface made up of electret grains. Now imagine another balloon within the first one, so that you have two balloons creating inward pressure. The pressure within the inner balloon would not be very great, but it would be more than the pressure created by the outer balloon alone. Now imagine a whole series of such concentric balloons, where the "rubber" of each balloon is stretched over the one inside it. The aggregate containment pressure of such an arrangement would be much greater than for a single balloon layer. We can see that, in a cloud of our dust grains held together by the slight pull of Van der Waals forces, the combined force of all the layers of lateral cohesion results in extreme pressure being exerted on the electrets toward the center of the cloud, greater than the pressure toward the periphery. This is not significant in smaller clouds but, as we will see, it becomes quite important if the cloud is large—very large.

Imagine an enormous electret cloud forming in deep space, far away from any stars or planets. Though removed from the influence of massive celestial bodies, this space can still be expected to contain cosmic rays flying in all directions and at various velocities, from just fast enough to strip atoms of their electrons to so fast that if these high-energy nuclei (mostly single protons) were to fly into our earth, they would penetrate a mile or so deep before being stopped. We can expect also that there would be dust grains in these regions of space, the flotsam of supernovas or other sources of ejected matter. We could expect some or even most of these dust particles to be electrets, with their electron halos extending outward, away from the center of the grain. With all these conditions in place, we can expect that there would be clouds of these electrets of various sizes. The most important thing about these clouds is that they would continue to grow much, much larger. In time, a cloud might attain a mass as great as that of our solar system.

The space between stars is so great that a giant electret cloud would probably form at considerable distance from the nearest star, although it might also be fairly close, as interstellar distances go. I do not know if there is a limit to how large such a cloud could grow. A very large one would exhibit some interesting properties. From the concentric-balloon scenario, we know that the pressure at the center of such a cloud would be very great indeed. The pressure would at some point become so great that the electron halos around grains of dust would be pushed aside, so that the dust particles would actually touch. At some point, in fact, these dust grains whose electron halos had been momentarily displaced would combine like bubbles to become one larger grain—one larger electret, that is, with its own combined halo. In the new, larger electret, as in the grains that formed it, the inward pressure of electrons covering the outer surface would be greater than the expansive force of the protons at its core.

If this can happen to one cluster of grains, it can happen to others. In time, all the grains near the center of a large cloud would combine to become larger ones, and the center of the cloud would become a huge mass of outsized electret grains. The smaller grains above the center would still be subject to significant pressure but would have to wait for the cloud to gather more electrets at its periphery in order to combine with other grains. You will remember that when electrons surround a grain to form an electret, their inward pressure is proportional to the charge at the core. When two grains combine, the positive charge on the new, larger grain now contains the combined charge of both of the original grains. The displaced electrons of the original electrets reassemble at the surface of the larger grain (in a multilevel arrangement, per the exclusion principle), and this covering presses inward on the core with sufficient force to ensure stability.

Now think of this as a process that repeats. Grains at the center combine whenever the pressure increases sufficiently on account of new grains falling into the cloud at its periphery. Eventually, the central region of the cloud becomes a huge electret mass surrounded by a still very large cloud of separate electret grains. The solid core will continue to grow as dust particles from further out are forced inward and merge with it. Their electrons join the others at the core's surface to press inward. The core is composed of an accumulation of small rock dust and contains within its growing mass the same proportion of radioactive elements as would be found in ordinary rocks on earth, possibly more. As it grows larger, the core gradually heats up from the radioactive decay of these elements. Of course, this all takes a staggering amount of time, yet there is plenty of that, as you will see further along in our story.

So far, we've been envisioning a large cloud of electrets with a massive electret at its center, composed of positively charged rock that contains some naturally occurring radioactive atoms. The latter continue to decay, generating more and more heat. Whenever electrets combine with one another to join the central one, they do not discharge but rather contribute their charges to the growing core. You can understand that at this point the sphere is both hot and positively charged, and both its temperature and internal positive charge continue to rise as the cloud surrounding it grows even larger.

At some point, the center mass of the new sun becomes incandescent, and from that point on the surrounding cloud can grow no more. The cloud must now *decrease* in size, even if additional particles fall within its perimeter, since now the central incandescent mass will neutralize a great many charged particles from the cloud to form a blanket of neutral matter around itself. This happens, because heat breaks down the electret condition, allowing halo electrons to move back inside the grains to form ordinary atoms with the embedded protons, and the resulting neutral dust forms a layer around the core. This layer continues to thicken.

The central incandescent mass is a hot electret compressed by its coating of electrons, lying beneath a layer of neutral matter. The individual electret grains that remain above the neutral layer are still subject to the inward pressure of the cloud and are pushed downward, but the integrity of the cloud is maintained by the repulsive force of halo electrons surrounding each grain. As some of these electret grains fall close to the glowing center, they are neutralized by heat and become normal matter. When this happens, those grains no longer merge with the electret core and do not add to its positive charge. In time, the blanket of neutral matter around the core will also become luminous by conduction and will assist in neutralizing and shrinking the cloud of electret particles above.

While all of this is going on at or near the core, further changes are taking place in the cloud of electret particles above. When the central mass first became luminous, the cloud above was essentially a great sphere, and its mass was roughly the mass of our solar system. The cloud was turning as the entire galaxy was turning—very slowly, but turning nonetheless. Now, as the cloud of electret particles begins to contract due to neutralization and deterioration close to the core, conservation of angular momentum causes the cloud to rotate more rapidly, and centrifugal force causes it to spread out and become flatter. The poles of the cloud as it turns pull inward, toward each other, and the equatorial portions expand. The cloud's deterioration continues, causing it to spin faster and faster. Eventually, it becomes a thick disk.

You will remember that Van der Waals forces are holding the cloud together. Their strength has a limit, and when that limit is reached, as the cloud shrinks and turns faster, its outermost mass of dust must detach from the rest of the cloud to become a more slowly turning ring. This ring is effectively in orbit around the inner cloud, which at this point is turning faster than the outer ring yet not quite fast enough to detach and enter an orbit of its own. As the cloud continues to deteriorate at its center and spins faster, it loses successive outer portions as rings, one at a time. Even the very last, innermost portion of the original cloud cannot maintain its position as the speed of rotation increases, until finally the entire cloud has been flung off as a set of concentric rings orbiting the central mass. The end of this process comes only when the interior ring is turning fast enough to achieve its orbit and pulls far enough away from the center that it no longer loses electret dust grains to neutralization and assimilation into the glowing neutral layer below.

Van der Waals forces hold each ring in its toroidal shape, so that all its particles move together. It is significant that each ring moves as a unit, that its particles all rotate at the same rate. The outer rings take longer to rotate than the inner rings. This means that the particles at the outside edge of an inner ring are moving at a different speed (i.e., faster) than the inner particles of the ring outside it, despite the fact that the rings are almost touching. That the rings cohere as units and rotate at different speeds has great implications for the eventual formation of planets, including ours. It will turn out that planets form between rings surrounding what will become a star.

It may be helpful to describe the entity we are left with at this stage: an electret cloud surrounding a massive core (the new sun) flattened and spread out to become a disk, the disk separated into rings. The rings are toroidal; they almost touch one another but rotate at different rates, just as our planets orbit at different rates about our sun. The outermost ring takes the most time to rotate about the central mass, as with our planets. The mass of the entire system we are picturing is nearly the same as that of the solar system it will become.

The ring system has much more angular momentum than does the central mass. The central mass, which will become the sun, has gained its angular momentum from the slowest-moving inner ring particles as they deteriorated and impacted its surface at some small angle. The central mass is turning, because the neutral particles that fell into it were rotating at a much lower rate than the particles that formed the rings. These impacts had to nudge the slower-moving central body and bring it up to speed.

So far, I've been describing the formation of our sun and a set of rings that will become the rest of the solar system. I wanted to get the sun formed first, so to speak, because it must have formed differently from what the accepted theory describes, in order to be consistent with an expanding earth.

I assume at this point that the rings surrounding our sun are almost touching one another. The dust particles in each ring move together, more or less as a unit. Of course, the rings are not a solid, but Van der Waals forces operate on particles of each ring, as they did on the particles of the cloud they came from, to keep them clustered together.

Now consider the electrons pressing in on the center of each electret grain. These thin out at the periphery, and their attraction to the center weakens, making it possible for them to be pushed aside when dust grains collide, as they sometimes will along the boundary between rings. There is a considerable difference in the relative velocities of particles in two different rings, so that when a particle from any given ring happens to collide with a particle from an adjacent ring, the collision occurs with such force that the two particles combine to become a larger electret, and the electrons that once surrounded the grains individually now surround the larger mass. This is the electret combination process, like the merging of bubbles, that I described earlier. Here, as before, the process takes place many billions of times, except that in this case an electret mass of combined grains forms *between* rings and grows very large. The process continues until a planet is fully formed, when the halo electrons pressing in on this mass eventually resist being pushed out of the way.

From the standpoint of our solar system as a whole, the concentric rings that break away into orbits about the sun produce large solid masses between them, and these become the planets. But a related pattern also occurs during a planet's formation to produce the solid masses we call moons. Our earth forms as a large electret between a pair of rings surrounding the nascent sun. The earth's halo extends outward into the surrounding space, so that the halo electrons of those particles that still remain in the adjacent rings will be swept up by the earth's halo electrons. In this way, the ring's particles are gradually gathered in by the earth's electrons to form a cloud of charged dust surrounding the new planet.

I should mention briefly that, at the outskirts of our forming solar system, the sun's outermost rings do not quite progress to become solid masses like the earth, and at this early point they remain charged electrets. For the moment, I will not write more about these outer planets, our "gas giants" and "ice giants," but will instead concentrate on our own planet.

The material that makes up the earth's electret was once the same charged dust that made its way into the sun's rings, and it is the same material that occupies the electret cloud above the new earth. Those particles continue to contain radioactive elements like those that heated the new sun and that today can be found in terrestrial rocks. It takes a long time, but the decay of those radioisotopes continues to heat the earth's electret from the inside. When this core becomes very hot, the new planet radiates heat into the cloud of electret dust above, neutralizing it. The neutralized dust then falls to the surface of the electret planet to form a blanket. This new surface, in turn, is heated by the core. It begins to glow with heat and neutralizes more dust from the cloud above. This sequence repeats for a very long time, until a great many of the electret particles have been neutralized and have fallen to the surface of the small earth. The remaining non-neutralized particles surrounding the new planet will, I believe, become our moon.

This neutral layering, as you may have guessed, is the same process that took place on the sun before the solar rings had formed. Gradually, the neutral surface thickens until a depth of more than twenty miles (thirty-two kilometers) has fallen to the surface of the small electret earth. This becomes a layer of hot, neutral rock. Because the layer is liquid at first, it is able to spread evenly over the whole surface of the new planet. I suspect the reason not all of the cloud is "brought down to earth" in this way is that electrons surrounding the earth's electret core do not extend upward through the neutral layer far enough to interfere with all of the electrons in the electret cloud above.

Although composed of molten rock, the new neutral layer insulates the core, sealing it in. At some point, the confined electret becomes so hot that it begins to break *itself* down. More and more of its constituent matter is neutralized and escapes to form liquid rock. This ordinary matter takes up more space than the electret matter it came from. It is crucial to realize that neutralized material released from an electret *expands*, since it is no longer being compressed by an electron halo. The resulting matter is now free to mix with the layer of neutral dust that fell earlier. Many more years pass, and as the electret gradually breaks down, the expansion due to internal neutralization continues.

This, in a nutshell, is the process that causes the earth to expand. As the core electret gives up more of its charge, less and less heat is required to make it discharge further. At this point in our story, the geological moment we may be most interested to hear about has not quite arrived, namely, the time when the temperature at the planet's surface finally becomes low enough that its rock crust solidifies. (The oldest rock we find on present-day earth dates back to that occurrence, some four and a half billion years ago.) Terrestrial rock remains a mélange of space dust and elements released from the earth's discharging electret.) This mixture—still molten as we consider it now—will in time spread evenly over the planet and become the earth's crust, and later our continents.

But again, I get ahead of myself.

Chapter 3 The Development of a Young Planet

We left the earth at the point where the surface is nearly cool enough to solidify, and the electret center of the new planet has started to deteriorate. This is a good time to write about the process of deterioration in the electret and its effect on the planet as a whole. I've said that the core of the early earth is a charged electret surrounded by electrons that maintain its great density, and that it has been formed from many electret dust particles that were originally between two of the rings around the sun. The process of deterioration is endothermic, in that it requires heat in order to go forward. The heat comes from the decay of the radioactive elements contained within the electret. As the temperature of the electret increases, its molecules move more rapidly. Once in a while, the most energetic atom at the electret's surface breaks free, despite the tremendous pressure bearing down on it.

The earth's electret, at this point, lies beneath the molten rock that is the crust. A detail crucial for earth's future development is that the elements within this liquid rock have spread evenly over the surface of the electret below. The earth's mantle has not yet formed, so the atoms that have broken through the core's surface and reverted to normal matter are free to become part of the molten crust. Whenever the electret loses a highly energetic portion of itself, it also loses a bit of its positive charge, since the atom that escaped was charged (that is, consisted of more protons than electrons) and now is instantly neutralized using one or more of the electrons from the electret's halo. The resulting atom is larger than it was before, and so the earth also gets larger—the earth expands.

This process of deterioration continues, and atoms freed from the core continue to mix with the molten rock above. All the naturally occurring elements are contained within the electret, which continues to lose matter. The crust above gets stretched as the planetary core it covers gets larger. At this point, another layer—the mantle—has formed beneath the crust but has yet to solidify. As the crust and mantle get thicker, they offer more thermal insulation, and the mantle only gradually cools. The electret below constantly loses charge, and the temperature required to discharge atoms as normal matter decreases. Eventually, the mantle does solidify and gradually becomes rigid, and a liquid metal layer develops beneath the now-solid mantle. The electret below continues to give up elements and in the process becomes cooler than the mantle above. The released atoms take up more space, forcing the new mantle to stretch. For some time, it does stretch, and the solid crust above it is forced to stretch as well, to accommodate the expansion below. This cannot go on indefinitely. When the mantle becomes too rigid, it must break open or crack.

The crack in the mantle takes place beneath the unbroken crust that completely covers the earth. I have spent many years trying to determine where this crack first started to open beneath the crust. Now I am not sure it is important. The important thing about the cracking of the mantle is that it took place beneath a continuous covering of crust somewhere in what is now the Pacific Ocean, long before the oceans formed.

As the mantle pulls apart, it changes the crust's surface. For many years, the crack lengthens and develops into much the same shape as the mid-ocean ridge we know today. Indeed, the mid-ocean ridge is the "scar tissue" covering the healed crack in the mantle that extends around the globe. Even as it lies beneath a still-unbroken crust, the earth is expanding, and the electret at its core continues to lose elements. The newly released elements continue to make the mantle thicker, even after the rift has opened up and begun to spread. Since gravity would tend to settle the mantle into an even thickness, liquid mantle elements that have just escaped from the electret core flow under pressure into the widening crack where they rise and fill the gap in the mantle's surface. These mobile materials carry heat from far below that renders the area at the top of the crack far hotter than the rest of the mantle. The rising material is at first more viscous than the rest of the mantle, so that when the mantle cracks again it does so at or near where it cracked before. The mantle separates, and the rock that has risen into the crack cools and hardens so as to increase the surface area of the mantle. The mantle thus grows out laterally from the crack line and continues to do so for many years.

The rate of the earth's expansion depends on the rate of its core electret's deterioration, and during this period that deterioration gradually increases to a level beyond even the current rate, driven by heat from two sources. At first all the requisite heat comes from decay of radioactive elements within the electret core. Over time, however, more and more of it comes from the hot mantle above. Even after it has cracked, the mantle continues to thicken, radiating more and more heat.

It's worth noting that, when the mantle was first formed, the electret was hotter, and the elements that escaped it were therefore hotter. At some point after the creation of the mantle, heat began to travel by convection back down from the hot mantle, through the liquid outer core to the surface of the electret.

Let us sum up the state of the earth as it exists today. It is covered by the crust, consisting of the continents and the ocean floor. Between the crust and the mantle lies a low-viscosity layer of lighter elements that becomes very important as the continents are pulled apart over the mantle. The mantle itself extends downward some 29,000 kilometers to the surface of the liquid metal part of the earth's core. Below this liquid layer resides the solid, charged, still compact electret core. This is the solid center of the earth that we know exists, but it is not made of iron or any other metal, or of solid hydrogen, as has been suggested. It is instead a dense electret made up of all the elements found in cosmic dust. It is all that is left of the much larger electret that formed the original, much smaller and denser earth.

An additional, surprising detail needs to be mentioned. Between the inner (solid) and outer (liquid) core lies a previously unknown layer. The electrons that surround the solid electret press up as well as down and form their own layer just on the outside of the inner core. This is a very interesting thin layer composed only of electrons. In order to press in to compress the inner core, the electrons are so close together that no metal atom of the outer core can invade their space. They form a region filled only with electrons that separates the inner from the outer core. The fact that this buffer exists makes possible an important and unexpected discovery: a satisfactory explanation of how the earth's magnetic field can reverse. I'll say more about that later.

Processes That Took Place Within the New Earth

Here, for easy reference, is a list of processes and milestones in the development of the young earth.

1. The planet was at first a highly charged, non-conducting electret body compressed by electrons into a dense planet much smaller than the present earth. It contained all the elements that now make up the earth, and it had nearly the same mass.
2. The cloud of tiny electret dust grains that surrounded the original electret was neutralized by heat due to decay of radioactive elements within the central electret. The earth's crust formed as these neutralized particles fell to the electret's surface. This layer became a molten mass that spread completely around the planet. At this point, there were only two layers: the electret core and an outer layer of molten rock.
3. The temperature of the earth's electret continued to rise, still from decay of radioactive elements.
4. The electret core began to break down and release its elements, which, having been neutralized, expanded as they escaped. This process expanded the earth.
5. The electret released more metals than non-metals.
6. All non-metals as well as metals released at this time were absorbed into the molten crust.

The Development of a Young Planet

7. The crust provided greater insulation as it increased in thickness. The electret below cooled as it decayed and lost charge.
8. The crust eventually solidified when its temperature got low enough.
9. When the crust solidified, the mantle that formed underneath was made up of neutralized dust and elements from electret decay.
10. The lighter elements worked their way up through the molten mantle to form a layer of these elements just beneath the crust. Later, these elements would act as a lubricant, allowing the mantle to move below the crust.
11. The mantle solidified, since as the electret lost charge it cooled down.
12. The heavier metallic elements formed a molten layer between the mantle and the electret. This layer increased in thickness to become the outer core.
13. Both mantle and crust stretched as the earth's electret continued to lose elements, which expanded as they were released.
14. The mantle stretched while still molten but eventually solidified; it was still forced to stretch for some time even as a solid.
15. The mantle eventually cracked and broke apart beneath the unbroken crust, which completely covered the earth.

16. The crack had much the same shape as the present mid-ocean ridge, and it was at this point entirely beneath the crust.
17. The rate of core deterioration increased because of heat that was now coming less from the decay of radioactive elements in the electret and more from the older mantle that had formed earlier at a higher temperature.
18. The electrons that pressed down on the solid electret now also pressed up on the liquid metal portion of the core as well, forming a low-viscosity layer of electrons between the two. This layer essentially disengaged the inner from the outer core, acting as a lubricant so that the inner core could spin with very little friction.
19. The electrons in this intermediate region also extended upward into spaces between atoms of the liquid core and thereby reduced its density near the boundary.
20. The electret at the center of the earth was by this point, and remains, composed not of solid hydrogen or iron or any other combination of metals. Rather, it contains the same assortment of elements found throughout the earth, except that in the core these elements are charged and compressed to an extreme density.

The Development of a Young Planet

Changes to the Inside Affect the Outside

Later in this chapter, I'll talk about how the crack in the mantle eventually broke the continuous crust apart to form the oceans and continents of today. But before we get to that, I want to describe the dramatic changes that the crust underwent during the many years that it remained intact and was stretched rather than broken by the earth's expansion. You may remember that the material that rose through the crack in the mantle was extremely hot matter that had escaped from the decaying electret shortly before. Those liberated atoms flowed beneath the solid mantle to the crack, where they rose to fill the gap. This hot material in turn heated the crust above the crack, enabling it to stretch rather than break apart. Evidence of this line of superheated crust is most apparent at the so-called "geosynclines" all around the world, which present the most obvious surface alteration caused by activity inside the earth.

Many mountain ranges are geosynclines and were directly created by the spreading of the mantle crack, which you can imagine as follows. As mantle separates under the crust, its crack tugs at the crust, increasing its area. This broadening must be taking place throughout the Paleozoic Era, since there are as yet no oceans. The crust stretches toward the areas that will become coastline, though at this point the continents have not divided, the oceans have not yet formed, and there is no coast. Later, the widening mantle will separate the continuous crust into segments and ultimately into the various continents. For now, though, the mantle remains beneath the continuous crust, and the crack drags the crust toward what will become the coast. Note that this pattern isn't universal. It's really not true of the Ural Mountains' geosynclines, because that portion of the crust was pulled north with the rest of Asia, *away* from the crack that later opened to form the Indian Ocean. The Ural Mountains' geosyncline was once over the crack in the mantle that later formed the Indian Ocean, long before that same crack led to the formation of the Himalaya Mountains.

But we are still at a time before continents and oceans. The crack is caused by the expansion of the earth; as the earth gets larger, the crack must get longer. Rather than splitting apart at this point; the crust stretches like taffy where it is heated at the mantle crack. It stretches slowly, allowing time for rain to wash nearby material—loose rock from erosion of the crust—into the depression that formed in the crust above the crack. The surface depression always fills in and remains almost level with the area around it. Over the course of many years, material from a greater distance away has time to wash into the depression. Over time the depression recurs, always above the crack in the mantle, as the continuous crust increases in area on both sides above the crack. In the Cambrian Period, I am told, the surface of the crust descends some

The Development of a Young Planet

forty thousand feet, or twelve thousand two hundred meters, in some places. The opening of the mantle under the continuous crust occurs throughout the Paleozoic. At the end of the Permian Period, in what will become the Pacific Ocean, the crack opens west of North America at such a rate that it succeeds, finally, in breaking the crust apart. What has been happening throughout the Paleozoic, from the Cambrian to the Permian, is that activity due to the crack has created and left evidence of intervals in geological time. Correspondingly, we find rock from the various periods in order, next to one another on the earth's surface. All of this, because the smaller earth expanded and caused the crack in the mantle to increase in length.

During the Paleozoic, there is water on the earth's surface. It escaped from volcanos before the Cambrian and evaporated. As a result, there is rain to cause erosion. Before the development of the oceans, this rain, as we now know, formed shallow seas atop the earth's continuous crust. My guess as to the depth would be about six feet, or two meters. Life on earth during this time was limited to shallow water.

I am trying to be brief, but here I should mention a few facts I discovered when I first realized how water accumulated on the earth's surface during the Paleozoic. At the end of the Permian period, there was a glacier that extended over the southern part of the earth's surface. There is ample evidence of this glacier, but not so much is written about its effect. There were as yet no oceans. All of the water on earth fell as snow onto that glacier. It extended far to the north where it melted to form a terminal sea where only bivalves that could withstand the cold could survive. All life that could not adapt to cold died out in one of the earth's great extinctions. Animals situated near that terminal lake survived, while most other life died off in the absence of plentiful water. It was not that life left the water so much as that water abandoned life, except in and around that cold lake.

Today, we are very interested in oil. If our geologists knew that oil and coal formed when there were no oceans, perhaps investigators would look for those resources in the correct places. There is little or no oil off the east coast of the U.S. There is some gas off the Nova Scotia coast, and a little oil. Oil was formed on the lower part of the North American continent as it was pulled away from South America, at a time when the crust completely covered the small earth. Actually, it formed in a large pool on the continuous crust extending over what is now Venezuela, the bulge of Central America, and up as far as what is now Oklahoma. This all happened during the Paleocene, before North America was pulled north and west, away from South America.

Now let me say a bit about the Appalachian mountain chain. Before the Cambrian, the crack in the mantle started beneath the solid crust and pulled apart for many years after it started. The crack lay beneath the part of the crust that would become the Appalachians in the United

States. It was discovered many years ago that the surface of the crust first formed a depression, which was filled almost immediately by surrounding detritus. What history leaves out is that it was heated from below by very hot liquid from within the mantle that rose to fill the crack in the mantle's solid top layer. As I noted earlier, the mantle atoms that first escaped from the electret rose through the liquid layer to the bottom of the mantle and then worked their way over to the crack, then up through it to emerge under the crust. The crust, in this case, was to become the Appalachian mountain chain. The rock at the bottom of the depression, heated from below, became so hot that it melted and rose, pushing up rock at the tops of the mountains. First, the depression filled with detritus from both east and west. Then, when the crack (and its heat from below) moved to the east, the rock-filled depression rose and was worn away. We must remember that events I've described in just a few sentences took place over millions of years. You will remember that so far there are no oceans.

We'll discuss this in greater detail later on; for now, it's important that the opening of the mantle stretched the crust throughout the Paleozoic and did not break it apart until the end of that era, so that all this stretching formed the periods we now know as the Cambrian, Ordovician, Silurian, Devonian, Carboniferous, and Permian. It was not until after the Permian that the crack in the mantle broke the solid crust apart to form the first ocean.

I do not know the exact shape of the curve that could be drawn to show the rate of expansion of the earth. It is due to the rate of decay of the electret that is now at the center of the earth as its solid core. The height of the curve would represent the rate of expansion, and the curve would progress (to the right along the time axis) for millions of years. The curve must start at zero since the earth would start to expand only once the electret started to lose its elements. It would rise very slowly at first, since the necessary heat would come only from the natural decay of radioactive elements. The curve would increase in height to some point during the Cenozoic, then fall until it represented the present rate. The present rate can be calculated. We can measure the separation along the mid-ocean ridges and estimate any stretching of the continents.

The earth's expansion accelerated because heat from the mantle, which had been formed earlier out of superheated elements released from the core, bore down on the electret. When the mantle crack began to open so quickly that depressions in the crust could no longer be filled in, the continents began to pull apart. I do not know exactly when this took place, but it was probably near the beginning of the Triassic period—definitely after the Paleozoic Era. I say this because we have Triassic rock laid down in the eastern part of North America.

This would indicate that the crack in the mantle extended up into the North Atlantic later than when it split Asia from North America. My guess is that the crack in

The Development of a Young Planet

the mantle started east of what is now the northern Philippines. It progressed northward and southward, and started to pull the crust apart to open the Pacific Ocean. The segment consisting of Asia, Africa, and Australia was still intact, and I am quite sure the split that created the Indian Ocean broke away and went up into what is now the Red Sea, long after the crack had extended up into the Atlantic. I say this because the Ural geosynclines indicate that Asia and Africa had to have been pulled north, taking the Urals with them.

It may be helpful to describe on a grand scale how the continents first broke apart. The Americas, North and South, were first connected directly with Antarctica in one large segment of crust. Asia, Africa, and Australia were also all together as a different segment, which was broken apart when it split and created the Indian Ocean. A consequence of this is that India did not scoot across the Indian Ocean and push up the Himalaya Mountains, as is commonly thought.

There is much work to be done to determine when all these events took place. I do not know, for instance, when the line of guyots in the Pacific that end with the Hawaiian Islands bent at a sharp angle. I suspect that that turn took place at the same time the Atlantic opened. The Pacific was being formed when the first part of that line of guyots was forming. Perhaps we could determine the age of the guyots and from this get an estimate of when the Atlantic opened.

I also do not know the exact size of the electret that constituted the original whole earth. I do know that at first the earth had no crust. When the cloud of charged dust had fallen to the earth's surface, but before the internal electret had started to deteriorate, the earth was at its smallest, yet with nearly all its present mass. I suspect it was less—perhaps much less—than half as large as it is now, but that is a guess. We do know how large the earth is now, of course, so we have one point on the earth's growth curve.

A second point could be established by adding together the areas of today's continents, islands, and shelves. After the continents had pulled apart, we would not expect their sizes to change very much, so the sum of their areas, in addition to the islands and shelves, would give us the total area of the original, continuous crust at the point when it had stretched as much as it could and finally cracked and began pulling apart.

There are obvious changes to the flora and fauna due to the earth's expansion, but I have been concentrating on geological changes, since I know more about the physical geology of our planet than about its animals and vegetation. Still, the story of life on earth should not be neglected in filling out the complete picture of its history. (I am, after all, an animal myself.) According to all we are told, there seems to have been very little life on earth before the Cambrian. Life in a great variety of forms is believed to have developed suddenly during this period. That does not sit well with me. I don't believe it.

I think that life developed over a longer period of time. There were pools of water on the crust before the Cambrian, and there would have been a great deal of time for life forms to develop in those early seas over perhaps a billion years anywhere on earth. When the first mantle crack caused the first depression in the crust, during the Cambrian, water would have flowed into that depression from elsewhere on earth. Because the mantle crack had only just begun to stretch the earth's crust, there would have been no significant mountains to interfere with water flow from anywhere. I suspect that many multi-celled forms of life that originated prior to the Cambrian flowed with that early water into the earliest Cambrian depression and became concentrated there, leaving a fossil record that gives the impression of an explosion of new life-forms during that period. I have no idea where the earlier pool of life would have been, but it does give us something to work on.

A few points should be made about ocean formation that have not been covered. The Pacific Ocean no doubt opened first, but here I will discuss what happened long after that opening. I will start at the point when the Pacific had been forming for many millions of years, up to the point when the Atlantic started to open. I expect you know by now that I think there is very little support for the idea of subduction. Subduction is the term used to describe the downward motion of one tectonic plate underneath another. The earth's expansion, as I conceive it, would cause the pulling back of the Asian continent off the ocean floor. This would mean that the mantle would *rise* gradually from under the continent, in a process I have called *eduction*. This scenario runs counter to the accepted thinking, of course. There is good evidence and interpretation developed by Benioff showing that the angle of the motion that results in earthquakes below ocean trenches is about 45°. But geologists know that it is nearly impossible to determine from seismographic readings the *initial direction* of that motion, and so they must use related knowledge to infer that critical piece of information. In this case, everyone "knew" (incorrectly, I think) that the size of the earth was constant and that the Pacific Ocean had always been there. It is perfectly understandable that they would all assume the first motion was down and not up. I believe, instead, that the ocean floor came up, and the land near the coast went down.

One of the great geologists of the past was the Dutch geophysicist Felix Andries Vening Meinesz. Working before the invention of modern instruments, he devised an accurate method, using pendulums, for measuring gravity at sea. Vening Meinesz did his experiments in a submarine where he was freed from the ever-present jostling of ocean waves. He discovered regions of lower gravity over the ocean trenches at the edge of the continental shelf and concluded that the continent must be holding the ocean floor down. I know now that Vening Meinesz had found evidence of eduction. His data were ignored.

Until the oceans were formed by the pulling apart of land masses (due, in turn, to the cracking and filling of the mantle), the earth was still quite small, covered by a continuous crust that would later become our continents and islands. The rift in the crust opened very rapidly in the Pacific, geologically speaking, and material that tore away from the continent of Asia is still present out in the Pacific, in the form of a rise in the ocean floor. It is my belief that if you drill into the Shatskiy Rise, the rock will be detritus from Asia and not the normal sea floor.

The deep trenches around the edges of the Pacific Ocean are the result of the earth's expansion. As the ocean floor pulls back from the continent, a gap opens in some places at the juncture between the two. There is only one trench that does not fit this pattern: the Puerto Rico Trench. While it *is* the result of expansion, the trench was created by North America's being pulled away from South America, and this makes it a little different. The divergence of North and South America was so rapid (in geologic terms) that the floor of the Caribbean could not stretch quickly enough to accommodate it. The result was a tear in the ocean floor that formed the Puerto Rico Trench.

As North America was pulled to the north and west, Alaska was pulled out, away from the rest of the continent.

This makes the Aleutian Islands a special case as well. This chain of islands is curved out toward the Pacific, showing that, as the Pacific floor was stretched, the islands were taken along for the ride. The rock is *greywacke*, the oldest of rock. The gravity there is less than it should be, indicating that the islands are lifted up somehow. My thought is that these islands are held above their natural level by the pulling of the ocean floor, because the earth is still expanding and the floor is stretching.

I've written about North America being pulled north and west by the earth's expansion. There have been times in my investigation when I've found answers to old questions that have very little to do with the expanding earth itself. One is the earthquake in the 1800s centered in Madrid, Missouri, in the middle of the North American continent where there was no apparent reason for an earthquake to occur. Yes, there was a fracture, but why the quake? My reasoning is that the continent is flattening, and as it does there is a change in the stress in the rock. Bending a slab of rock upwards at the edges produces a compressive stress in the upper part of the slab and a tensile stress in the lower part. I think the fact that the earth is gradually getting larger and therefore flatter creates that stress, and that stress was responsible for the earthquake in Missouri.

I wrote earlier about landmasses being pulled apart from the supercontinent comprising what are now North America, South America, and Antarctica, as it was being

The Development of a Young Planet

stretched by the earth's expansion. Two seas were created directly by this stretching. They are different in many ways. There is, however, an important similarity between the area known as the Caribbean Sea and the area of seas caused by the pulling south of Australia away from what had been the Asia-Africa-Australia segment. This is, of course, the sea surrounding the islands of Indonesia, from Java to Borneo, plus the Philippines. Both these seas are the result of stretching apart of continental segments. Each tract of ocean contains many islands that were torn away from their respective segments by the stretching of the sea floor.

Chapter 4   Electrets in the Cosmos

Light and the Red Shift

In an earlier chapter, we found that the dust in space can be charged by cosmic rays, then acquire a perfect sphere of electrons around itself. This could not happen on earth, only in the cold of deep space, where, after many hits by cosmic rays and in keeping with the rules of quantum mechanics, a halo of electrons will develop around the charged dust particle. Such an electret can persist only because it is in a cold environment. If it combines with other like particles to become a much larger mass, then its charge becomes very great, and a great deal more heat is required to neutralize it. As light approaches us from deep space, photons must pass though the halos of electrons that surround these electrets of various sizes. The light interacts with the electrons and in doing so loses energy, so that its wavelength is greater and nearer to the red end of

Electrets in the Cosmos

the spectrum. This "red shift" does tell us something about the nature of the universe, but not in the way we previously thought.

You have, no doubt, observed that when an object is partly submerged in water, the part under the surface seems to have moved. What has happened is that the speed of light reflected off the object is greater in air than it is in water, resulting in a change of direction for light passing through the heavier medium (refraction). Likewise, when a photon of light enters a halo of electrons, it is slowed because the halo has some slight density relative to the vacuum of space. After light goes through the halo and comes out, it goes back to the speed of light in a vacuum, but it has lost energy and so has a longer wavelength than when it went in. Certainly the change in velocity is small—actually, it is all but negligible—yet it is real, and after the light has gone through many billions of halos, it can lose enough energy so that the increase in its wavelength is measurable. That movement of the light's wavelength toward the red end of the spectrum is responsible for the red shift.

When light from distant galaxies goes through a spectrometer, it is broken into spectral lines. Edwin Hubble noticed that, for light from distant galaxies, the lines rendered by the spectrometer were shifted toward the red end of the spectrum and that the farther the galaxy was from earth, the greater the shift. Two years before Hubble made his discovery, Einstein had determined that light from objects moving away at relativistic velocities would be shifted to the red and that the greater the velocity, the greater the shift.

Hubble made the obvious deduction that the objects he was seeing must be moving away at tremendous velocities. Einstein was correct in his analysis. It is true that when objects are moving away at relativistic velocities, light coming from them is shifted to the red end of the spectrum, and when objects are moving toward us at these tremendous speeds it is shifted toward the blue. And Hubble was correct in his observation of the red shift, but I believe he was wrong in concluding that the galaxies he observed were actually moving away.

At the time, there *were* scientists who thought that perhaps the light had lost energy on the way to us through space. One was the Swiss astronomer Fritz Zwicky, who later made many scientific discoveries in physics and became quite well known. Zwicky came to the U.S. in 1925, the year I was born. He became a research fellow at the California Institute of Technology in Pasadena and eventually studied supernovas. As an astronomer, he found eighteen supernovas; up to that time, only about twelve had ever been recorded. Zwicky had suggested that perhaps dust in space caused light to lose energy as it traveled toward us, but this was not mentioned in his obituary. As far as I am concerned, that was his most important insight, because he was right. I was told that the reason light could not be slowed by dust was because if it lost energy by striking a dust particle, light would be diverted from its straight path, and then of course we would not see the object. But we do see the distant galaxies, so that idea cannot be right. It is true that a *collision* would alter the light's path, but that is not how the dust interferes with the speed of light. The photon loses energy to the electron halos around the dust particle, not by bouncing off the dust particle itself. Zwicky was right in his overall conjecture, but he didn't know about the halos and how they caused the loss of energy. Light does lose energy to dust particles—to the halos of electret particles, that is—and as a result its wavelength increases. With this explanation, major conclusions in astronomy and physics must be reversed: distant galaxies are not moving away at relativistic velocities, and the universe is not expanding, nor was there a Big Bang.

It is the present theory that if galaxies are moving away from some place in the universe, then that must be the place where the universe started. All of the universe was thought to be at, or contained in, that point, so that the universe started from this "singularity" with a tremendous explosion. Hoyle called the event the "Big Bang" as a derisive name, but it stuck. He fought against the idea for years. He was right all along, though he didn't know why. There was no need to speculate about a Big Bang in the first place, since light from distant sources is not redshifted because those sources are hurtling away from some central location, but rather because it loses energy to electret halos as it travels through space.

If the universe is ageless, how does it keep going on? We know that stars eventually die, so how are new ones created to take their place? The answer is that light gives the electrons in the halos a tiny push, and the electrons are attracted to the dust particles at the center of the halo, so that the dust is also given a tiny push in the direction the light is going. If that is so, then the dust is driven out of the galaxy in time—a very great time. I contend that all the stars in the galaxy will eventually explode as supernovas and turn back to dust, but this will take many billions of years. I suggest that there is time enough for all this to happen, since there was no Big Bang, and there will be no final resolution to the universe at the other end of such a process. Any galaxy would go through the same process, in which all of its stars go supernova. The result is that the stars return to dust, and the dust in time develops its own halo of electrons. Each particle, over a long time, is moved by light, propelled by the "sail" of its halo, and so is moved into the darkest regions between galaxies. If it is moved toward an existing galaxy, it is driven away by the light from that galaxy until it eventually accumulates in the emptiest regions in space. Ultimately, stars will form through the processes described in Chapter 2, and many stars will form new galaxies ad infinitum. (I asked my wife with the Latin major how to spell "ad infinitum." She was also the one who came up with the Latinate term "eduction," the opposite of subduction.)

We can't leave this topic without addressing (19th-century German astronomer Heinrich) Olber's Paradox. The paradox seems to arise under the assumption that the universe is without limit and has an infinite number of stars. If this were the case, Olber argued, the night sky should be completely filled with starlight, since, if we go far enough along any straight line from the earth, we should find a star acting as a light source. But the sky has dark patches where there is no visible starlight. The universe, therefore, cannot have an infinite number of stars. The paradox was well-answered by the Big Bang theory and the limited universe it implied. But, since I see the universe as unlimited for reasons I've given, Olber's Paradox will require a new answer. Light loses energy as it travels through space by passing through the electron halos of electret particles. Whenever this happens, the light's wavelength increases, and it continues to do so on repeated encounters with halos, until its wavelength becomes greater than the diameter of the halo, after which its wavelength is no longer affected. At this point, what was once visible light has become radio waves, which we *do* find emanating from every region of space. Actually, we might be able to determine the diameter of the halos by examining the radio waves that began their journey through space as visible light.

Usually, the red shift is a good measure of distance, although for different reasons than are usually given. But there are times when we are fooled into thinking an object is farther away than it actually is. You will remember that as our sun formed, its light emanated through a cloud of electret dust particles, each with its own electron halo. After traveling a comparatively short distance, the light of the new star passed through many billions of halos and lost energy. As a result, the wavelength of light emerging from the space around the nascent sun was much greater than it would have been after traveling the same distance elsewhere in space. To any observer beyond the young solar system, if they assumed that the wavelength of light emanating from it was a function of distance, our embryonic sun would have seemed much farther away than it actually was. So I assume that the thousands of red-shifted quasars we see today are not as far away as they are thought to be. They are just newly forming stars, most of them in our own galaxy. We assume they are far away, because we are accustomed to using the red shift as a measure of distance. This is one of those instances when doing so is misleading. Halton Arp lost his position as an esteemed astronomer at Mount Palomar because he kept pushing his idea that the quasars are not far away. He was right, but it seems I can't help. I wrote to Arp and offered my explanation for the red shift, but so much would have to change if I was right that he could not see things my way. I am convinced that the only people who can accept my ideas are those who are not deeply grounded in the current scientific consensus. The experts already know what has been taught and know it well.

Another example of our being fooled into thinking an object is farther away than it actually is has led to the idea that there are voids in space. Such voids do not exist. Instead, there are some volumes of space that contain a greater density of electrets than usual. Light moving through these electret clouds shifts more to the red than it would moving through other regions of space where the density of electrets is more typical. This pronounced shift to the red makes objects on the far side of these electret clouds appear to be farther away than they are, making the volume of space containing the cloud seem larger than it is. Thus, what appears to be a vast and empty void in space is actually a much smaller volume with an unusually great concentration of electrets and their halos.

In short, there is no need to posit a Big Bang to explain the red shift, since it can be explained entirely by the existence of electrets in space.

Supernovas

There is a bit more to say about the nature of the cosmos. So far, we have discussed the formation of the sun, the formation of rings around the sun, the formation of planets out of those rings, and the supposed voids in space. I've covered the fact that the universe has no size or age limit and described how particles of dust are continually driven out of galaxies to form new stars and, eventually, new galaxies, when those new stars become plentiful. That is quite a lot on the topic of deep space, but we have not yet discussed supernovas and comets.

Let's start with supernovas. I hinted that when the sun formed, its core had to have been under tremendous pressure from its surrounding electrons in order not to have broken down at the end of its accumulation of electret particles, before its system of rings had developed. If I am correct about all of this, then the sun's core will not break down, even though it is extremely hot—not, that is, until the sun finally explodes. When the pressure of the electron shell surrounding the sun's core is somehow disturbed so that it can no longer contain the explosive force within the core, the result will be the tremendous explosion that we call a supernova.

I'm using my understanding of how the sun formed to understand supernovas. Similar reasoning could also be used to explain novas, which I understand involve smaller explosions. Let me outline what I know now about the core of any star. The core of a star is like the inner core of the earth, in that it is held to great density by the shell of electrons pressing in on it. The difference is that the pressure on the sun's core is much greater, so great that it has never deteriorated in the way the earth's core has. Under the tremendous pressure of the electron shell surrounding the sun's core, the composition of atoms contained within it may have changed. The great pressure and heat may have caused atoms to form different elements, either through nuclear fusion or some heretofore unknown mechanism. Hoyle discovered the order of the formation of elements on the periodic table, up through iron. I am not going to try to increase the list by speculating as to how the higher elements formed, but I suspect they were formed within the core, because it is under much greater pressure than previously thought.

Perhaps the shell of electrons pressing in on the core of the sun will one day be disturbed somehow, so that inward pressure is not as great as outward. It may be that elements above the shell of electrons will invade the shell and force electrons away from their positions around the core. Then they will be unable to exert sufficient pressure on the core to hold it together. Once that inward pressure is released, the core, which generates tremendous outward pressure because of its incredibly large internal positive charge, will simply explode. All the matter inside will suddenly be free. As it flies through the surrounding electrons, the charged dust will pick up electrons and revert to normal matter. Some particles—specifically, the nuclei of individual atoms—may be moving too fast to pick up electrons and may hurtle through space as cosmic rays. This, then, is the outcome of a supernova: the disintegration of a star and its replacement by ordinary space dust and possibly cosmic rays. I think that happens to all stars eventually.

We still need to think about what could happen to the matter outside the shell that would force the electrons out, away from the core. When this material presses in on the electron shell with great force, the outer electrons may work their way up between the atoms above. It should be possible to determine how the outer part of an older star changes chemically so that it penetrates the electron shell enough that the force inward becomes slightly less than necessary to keep the star's core compressed.

Evidently, the inward pressure is only slightly greater than the outward. Certainly, it is possible that another star could strike the old star and that this would disrupt the electron shell and cause a supernova. Perhaps this is what happens. It could account for novas as well as supernovas. Similarly but on a smaller scale, if a heavy meteor were to fall into the old star without breaking up on the way in, it might disrupt the electron shell's ability to maintain its net inward pressure.

Comets

The present theory of comets can also be improved upon. I believe that there are at least three different kinds of comets. Comets can have a small solid center, as we found when the European Space Agency crashed its Rosetta spacecraft into a comet in 2016. Then there is the type that fell to earth in the Tunguska event of 1908, an explosion in the atmosphere above Siberia that knocked down trees for miles around. There was no meteorite or crater found, but there was a tremendous explosion. Finally, there is the kind of comet that resulted in the many obsidian-like rocks called tektites strewn over several parts of the earth, including North America, the Czech Republic, the Ivory Coast, Southeast Asia, and Australia.

I originally thought that comets had no core or nucleus, but my mind changed as my understanding grew. My first image was based on our old observations from here on earth, where we could not make out a nucleus at the center of the comet. I knew there were electrets in space and thought at first that comets were just an accumulation or cloud of these electrets and nothing more. I still believe that at least some of them are small clouds of electret particles, but others may consist of a cluster of larger electrets. When I was writing about the formation of the sun (see Chapter 2), I described how electrets at the center of a cloud could combine through a merging process. There must be objects in space that are at various stages of this process. Some comets may be a loose collection of such objects, each with its own electron halo.

The Tunguska event was likely caused by a collision with the earth of a comet of the former type—a cloud of tiny electret particles. Such a body would certainly explode and cause the kind of damage, along with the lack of any evidence of impact, seen in the forests of eastern Siberia. Tektite-producing comets seem to be of the latter kind—not clouds of tiny electret particles, but rather clusters of large electret "chunks." These chunks explode and form pieces of obsidian-like material that rain down on the earth's surface.

Chapter 5 Evidence Today of the Earth's Expansion

Expansion Created Geological Records of All the Periods of the Paleozoic

I briefly noted the development of the rock formations of the various periods during the Paleozoic, but I believe this topic deserves further discussion. Geologists have spent many years studying the various rock formations and life forms from the Cambrian to the Cretaceous. The entire earth was covered by a continuous crust all through the Paleozoic. The crust was stretched to different degrees in different places across the Cambrian, Ordovician, Silurian, Devonian, Carboniferous, and Permian periods, but because it was still unbroken, there were no oceans. Early life forms did develop underwater in this era, but they did so in water lying on the surface of the crust. The oceans we know today formed only after the crust cracked and the continents started drawing away from each other. Prior to that point, activity in and around the crack in the mantle, beneath the unbroken crust, defined the various periods of the Paleozoic—new information that I expect should be of interest to any geologist.

I was brought up in New York State, on Lake George in the Adirondack Mountains in the northeastern part of the Appalachians. I believe that the crack in the mantle progressed eastward, underneath the portion of the crust that is now North America, heating and stretching whichever portion of the crust happened to be above it at the time. The crack was beneath the Appalachians when that range developed. I read of the discovery by James Hall in the mid-19th century of how the geosynclines developed. This led to my understanding of how the rock strata that characterize the different periods in the Paleozoic were created. In 1893, Hall was appointed state geologist of New York. He walked across the state looking at the rocks and determined that the land first sank in a long depression and filled as it was depressed. I can add that material from both sides washed into the depression as the crust slowly went down. In the case of the Appalachians, the heat rising from the spreading mantle heated the crust along a line, so that continued pulling (caused by the earth's growth) stretched the still-unbroken crust. While it was still hot, the semi-molten rock of the crust stretched like taffy. This first caused the crust above the crack to become thinner and thus to sink. The crust did not stretch equally on either side of the mantle crack, so eventually the location of the crack shifted east relative to the crust and began heating and stretching the crust in a different location. The material that had washed into the original depression solidified and was then squeezed from the west by the gradual flattening of the crust. Caught between a rock and a hard place, so to speak, this material had nowhere to go but up.

Let me explain. As I noted in Chapter 3, when a concave slab is bent to make it flatter (as happens to any portion of the crust as the earth gets larger), the upper surface of the slab is subjected to compression, while the underside goes into tension. The slab in this case is the continental crust. Here we are using as an example mountains in the Appalachian range, which runs north and south, but the process holds for any range formed during this era, regardless of its orientation.

At a later point in time, a mantle crack beneath the western part of the North American continent caused another linear depression that was subsequently pushed up in the same way to create the Rocky Mountains. The Rockies were originally much higher than they are today but wore down to their present size over the years.

The same thing happened throughout the Paleozoic. In one location after another, a depression in the crust caused by a crack in the mantle beneath it was filled in by rock, and this rock was later compressed and thrust up after it had hardened, following the movement of the mantle crack to a different location. As a result, rock strata representing different geological periods can be found today in different regions of the surface, depending on where the mantle crack was located during that period.

Some time ago, when I was reviewing what was known about the early geosynclines, the subject of coal production came up. As you may know, much of the anthracite coal in the U.S. is mined in the Appalachians. The coal is deposited in layers called "coal measures," in which veins of coal are interleaved with layers of sedimentary rock. The reason for mentioning this is that the expanding earth explains how coal measures formed, where previously there was no good explanation.

As the earth's surface sank into the depressions formed by the crack in the mantle, swamps formed containing vegetation that eventually fell to the bottom of the depressions. As the earth expanded, the depressions deepened. Of course, the rock beneath the swamps did not accommodate this movement smoothly. Rather, pressure built up until the rock adjusted violently in the form of intermittent earthquakes, each of which was accompanied by a sudden drop of the swamp floor by several feet. Each time this happened, the depressions were then filled in, first by heavy rubble then by finer detritus washed in by falling water, until the swamp was again close to level. The depressions filled in rapidly, before much additional vegetation could accumulate on the swamp floor, because gravity was so much greater at that time that the water and rubble tore away anything in their path. Once the depression filled in, vegetation again accumulated which would eventually become a new seam of coal, separated from the one below it by the layer of sediment that had rushed in to fill the newly deepened depression.

The reason that heavier coal formed instead of lighter coal such as lignite was that the atmosphere held a higher percentage of oxygen at the time. Under high pressure and with higher oxygenation, the vegetation did not rot, because the oxygen killed off the microbes necessary for decay. When you pour sulfuric acid (a strong oxidizer) on sugar, you get black carbon. This is similar to the effect of the highly oxygenated air on the fallen vegetation in those primordial swamps.

The Original Continents Comprised Several of the Continents We Know Today

While all the rock formations of the Paleozoic were being created, the oceans had not yet formed. The Pacific Ocean formed before the Atlantic, which opened during the Mesozoic Era. I say this only because the rock of the Triassic, the first period of the Mesozoic, lies beneath the East Coast of the U.S., so that is where the crack in the mantle was located during this period, and that is where it was when the crust broke apart to begin forming the Atlantic Ocean.

It seems that the way continents formed was that, first, the earth's crust was broken into large segments, and then those segments were broken apart into smaller land masses by the earth's expansion. The initial two segments, or supercontinents, were joined in the north and were much longer when the earth was smaller. There was an American segment that included North America, Central America, South America, and Antarctica. The Asian segment was much larger; it included Europe, Asia, Africa and Australia. These segments were separated, as they remain today, by the crack in the mantle, which later became the mid-ocean ridge. As the earth became larger, the mantle crack also had to get longer. Today's so-called "transform faults" are evidence of this. The American segment stretched to break Antarctica away from South America. When North America was pulled north and west away from South America, Central America unfolded out of North America to form the Gulf of Mexico and the Caribbean. The evidence for this is that the Yucatan Peninsula can be folded neatly into the space between Florida and Mexico. I believe that at some point North and South America separated completely. At that point, what is now southern Nicaragua was connected to the northeastern part of South America. It was subsequently dragged west across the northern part of South America, pulling Panama and Costa Rica away from South America in the process. I do not know where the split occurred, but comparing the geology of Colombia and Venezuela to that of lower Central America should tell us.

Hess, Ewing, and the Age of the Oceans

If you look up Harry Hess on the internet, you can read all about him and his discoveries, some of which were very important to me. Hess was a professor of geology at Princeton University. I understand he joined the Navy right after the bombing of Pearl Harbor in 1941 and, because of his position and training, was given a small ship. He arranged for it to have the latest sonar equipment, which he kept running as he sailed back and forth across the Pacific. As he traveled, he found flat-topped mountains rising from the sea floor. He had discovered "guyots," which he named for 19th-century geologist Arnold Henry Guyot. It was later found that they had been volcanoes worn down by wave action, but some of these tablemounts were nearly 700 feet beneath the surface. Hess saw them as the result of his "seafloor spreading," but for me they provided evidence that the ocean's surface was once down at that low level. His notion of seafloor spreading developed into plate tectonics, making Hess the father of the current standard theory of movement in the earth's crust.

There are other geological discoveries that provide evidence that ocean levels were once much lower than they are today. Many years ago, my geology professor at RPI told our class about the Hudson Canyon, which had just been discovered. This submarine canyon was odd, in that it had been worn into the solid rock running from the continental shelf off New York to the deep floor of the ocean. He said this erosion was impossible to explain and hoped one of us could determine its cause. At the time, I had no idea that I would find the answer, but I now know that the water level of the oceans was much lower when the continents first pulled apart—thousands of feet lower than today, in fact. This means that it must have been the Hudson River running over the rock of the canyon, which was exposed at the time, that wore it down. Of course, there are many undersea canyons at the mouths of old rivers around the globe that can now be explained this way.

The only water in those early oceans was runoff from the primordial continental seas into the new openings in the crust as the continents broke apart. In time, as the crust continued to pull apart, the many volcanos around the world continued to give up water as steam. This became part of the atmosphere and eventually fell to the earth as rain, which in turn found its way into the spreading oceans, in time filling them. The process continues to this day, although until recently the opening of the ocean basins due to the earth's expansion has kept the ocean near its present level, even as additional water has found its way to the earth's surface.

In 1947, the Columbia University geologist Maurice Ewing was able to secure use of the Woods Hole Oceanographic Institution's research vessel Atlantis to search the ocean floor. In an article he wrote for the National Geographic, as I recall, Ewing described sandy beaches two miles beneath the surface, as though the ocean level had been that low millions of years earlier. He knew (or thought he did) that such beaches were impossible, yet still he described them.

All of this tells us that the continents at one time extended all the way out to the edge of what is now the continental shelf, with the surface of the ocean far below. Today, of course, the oceans have risen to cover the edges of those original continents, but in reckoning the size of the continents, we have to include the continental shelf. This becomes crucial in determining the size of the earth at the time the continents first began breaking apart. Up to that point, the entire earth was completely covered by a shell of continental crust. It is straightforward to calculate the size of the crust then, since we know the current sizes of the continents, along with their continental shelves, and of the islands that have been pulled away from them. The sum of all these together represents the size of the continuous covering of the earth a little over 200 million years ago. This was about when the crust was last intact, because 200-million-year-old ocean floor is the oldest that could be found by the Glomar Challenger deep-sea research and drilling vessel.

Since I mentioned earlier that I had spent quite a bit of time trying to determine where and when the crust first pulled apart, let me elaborate on how I think it happened. As I've said, the crust finally reached a point where it could no longer stretch, and so it broke. My opinion is that this first took place near what is now Japan, before Japan had broken away from the Asian continent. Material from the continent washed into that first crack in the crust to form a slope made up only of continental rock. That rock can be found today in the slope that extends from the edge of the continental shelf down to the ocean floor.

Later, the earth continued to expand, and the ocean widened. As this was happening, the crack got longer very slowly. The Pacific Ocean formed first, after which the rest of the crust remained solid for a long time. The fact that the Glomar found ocean floor 200 million years old and none older suggests that the crack in the crust must have formed only slightly earlier, after which water from the continental seas began to drain into the ocean basins. At first, the oceans were not filled, and their water level was lower—several hundred feet lower, when the guyots first formed as wave action wore down the tops of the original volcanos. The Atlantic's water level must have been lower yet, when the Hudson River carved what is now Hudson Canyon into solid rock out to the edge of North America's eastern continental shelf.

Evidence Today of the Earth's Expansion

Volcanoes and the Andesite Line

I am not going to go into the present theory of what happens as lava rises through volcanoes—that theory is correct—but rather will say what I think causes the formation of volcanoes in the first place. Most, though not all, volcanoes are found inland, off the trench that makes its way around the Pacific. In my view, the floor of the ocean rises up from under the continents rather than going down beneath them. This is the opposite of subduction theory. I believe that, where the mantle is rising from deep in the earth, its molten rock rises along the so-called Benioff zone (after its discoverer, Hugo Benioff of Caltech) at an angle of some 45 degrees from the bottom of the trench. As it rises from beneath the continent, the mantle breaks rock off the continent's edge. (Today, as noted earlier, a continent's edge is typically submerged and forms the continental shelf.) Hydrogen and oxygen trapped within the mantle combine when they are close enough to the surface to be released, and of course they try to escape. The edge of the continent that is holding these gases down is being pulled toward the trench and the ocean by the movement of the mantle beneath it. This pulling action naturally creates fissures and cracks in the rock, and this in turn creates an opportunity for gaseous materials to escape upward, lowering the gas pressure below. When this happens, the path is now open for water, in the form of steam, to push up and out through the fissures in what has become a volcano. Carbon atoms in the molten rock pick up some of the oxygen from the steam to form carbon dioxide, and this gas, along with the nitrogen and water, emerges with the rock. As these hot gases rise, they melt continental rock along the sides of the opening, and pieces of detached rock and dust fly out of these volcanoes.

We find that there are two types of volcanoes: the kind found on the landward side of an ocean trench, which I'm focusing on here, and another, less common kind that rises directly out of the mantle itself. In the Western Pacific, a geological border known as the "andesite line" runs parallel to the deep trenches in the Pacific Basin. Volcanoes on the landward side of the andesite line contain continental rock mixed with mantle rock, rather than the mantle-only type of rock found in Hawaii. The mantle (as sea floor) is being stretched, and this must be the cause of the mantle-only volcanoes that come up from the seafloor. This type is typically less explosive, since gas and lava do not need to break through continental rock to reach the surface. The explosive type of volcano, on the other hand, is the kind found inland from the andesite line. Here, the weight of continental rock holds the magma and gases beneath it under greater pressure than the mantle rock of the sea floor would. When those gases escape, they drive dust and rock skyward, sometimes in the classic "plume." The famous Krakatoa eruption was of this type, as was that of Mount St. Helens.

The reason I'm writing about these volcanoes is that they emerge from under continental-type rock, where ocean floor is moving seaward because of stretching due to the earth's expansion. The pull the ocean floor exerts on the edges of continents causes stretching and tearing of the continental rock, which in turn creates the fissures that allow for the formation of volcanoes. Some of the deepest trenches and the most active volcanoes are the result of ocean floor being pulled out from under continental shelves.

What about Mount St. Helens? When North America was pulled north and west by the rending of the American-Antarctic supercontinent, some of the continental rock was cracked by the pulling, and mantle gases escaped to form volcanoes such as St. Helens. The same is true for volcanoes in Italy, because the Mediterranean Sea was opened up as the Asian-African-Australian segment was pulled apart and Africa moved southward.

Chapter 6 Other Mysteries Explained by the Earth's Expansion

The Earth's Magnetic Field

I said I would revisit the layer of electrons separating the outer and inner core of the earth. The electrons in this thin layer are pressing down on the surface of the inner core with tremendous pressure, and in order to do so they must be very close together, so close that they exclude all of the metal of the outer core. The layer is made up only of electrons and so contains no atoms. I therefore suspect it offers very little resistance to prevent the inner core from turning at a different speed than the liquid-metal outer core.

Because they are composed of atoms with some of their electrons missing, electrets have a strong internal positive charge. When positive charges rotate, they create a magnetic field. So when an electret, such as the inner core of the earth, rotates, it creates a magnetic field. But a positively charged object rotating in the direction that the earth is rotating would create a magnetic field whose polarity is the opposite of the field that actually exists at the Earth's surface. So why would the earth's magnetic field have the polarity it does if the inner core is positively charged? We need to remember that the inner core is surrounded by a layer of electrons, which of course have a negative charge. When a negative charge rotates, it too creates a magnetic field but one that is opposite in polarity to the charge created by the rotation of a positively charged object. When the shell of electrons surrounding the earth's core rotates in the direction the earth turns, it creates a field that matches the polarity of the earth's present magnetic field. The two fields, the one inside created by the rotation of the electret and the one outside created by the shell of electrons, partly cancel each other out, but the field generated by the rotation of the shell of electrons is stronger at present. We should not be surprised at that, since the absolute speed of electrons in the shell is faster than the absolute speed of particles in the electret core. That's simply because the electrons have farther to travel during any given rotation than do particles closer to the center. What's puzzling is that the polarity of the earth's magnetic field reverses itself every few hundred thousand years. Whereas the earth completes a rotation every twenty-four hours, the inner core electret may not rotate at exactly the same rate. The shell of electrons surrounding the inner core electret, too, may not rotate in sync with the surface of the earth. These statements will require some explanation.

As the earth expands, its rotation must slow slightly, in keeping with the conservation of angular momentum. Because the inner core is decoupled from the outer core by a layer of electrons, the inner core would not have to slow down just because the outer core does. I suspect there would be very little friction between the inner and outer core by virtue of the electron shell, which acts as a lubricant between the two. The inner core would keep going at nearly its old rate of rotation when the outer core slowed down. At the same time, the outer core, which we know to be liquid metal, would turn in tandem with the surface, but the electrons of the shell may not. The electron shell is pressing in on the solid electret, and so we must expect at least some of its electrons to move in sync with the inner core, but the outer part of the electron shell comes in contact with the liquid-metal outer core, and those electrons would move with the rest of the earth and slow down as the earth expands. At any given time, the magnetic field generated by the halo of electrons is proportional to the net speed of rotation of all those electrons taken together. That speed is somewhere between the speed of the inner core and that of the rest of the planet.

At present, the field created by the electrons is dominant, and the polarity of the field about the earth can be

Other Mysteries Explained by the Earth's Expansion 83

attributed to the rotation of the electron shell. With time, however, the earth will continue to expand, and the difference between the rotational speed of the electret core and the rest of the earth will become greater. The rotation of the electret core will also become increasingly faster than that of the electron shell, whose speed of rotation "splits the difference" between those of the inner and outer core. As the inner core moves faster relative to the electron shell, its magnetic field becomes stronger relative to that of the shell. It has been observed that the earth's magnetic field is now weakening. This indicates that the strength of the field generated by the electret core and that of the electron shell are balancing out. Eventually, the field generated by the core will dominate, at which point the polarity of the earth's magnetic field will reverse.

This explains how the polarity of the earth's magnetic field could reverse itself once, but according to the literature on the subject it reverses itself roughly every 200,000 to 250,000 years, with wide variation to that figure. Since it was at some time the opposite of what it is now, and since this was caused by a faster rotation of the inner core relative to the rest of the planet, something must have slowed the rotation of the inner core at some point in the past, bringing it closer to the speed of the rest of the earth. And since the earth's magnetic field has reversed itself not just once but multiple times, there must have been multiple occasions when the rotation of the earth's electret core was slowed, allowing the field of the electron shell to dominate. And then each time, as the earth further expanded, the rotational speed of the electret core became gradually faster than that of the rest of the planet, until its magnetic field became strong enough to overcome that of the electron shell, and the planet's magnetic field reversed back.

Under normal circumstances, the inner and outer core move through space as they go steadily around the Sun without interfering with each other, isolated from each other by that thin shell of electrons. But occasionally the earth is struck by meteors large enough to throw it slightly off course. We can understand how the inner core, which is not directly attached to the outer core and the rest of the planet, would attempt to continue in the direction it was going before the impact. The result would have to be that the inner core would rattle about within the outer core. This increased contact between the inner and outer core would slow the inner core, causing it to rotate at something closer to the speed at which the rest of the earth is turning, at least for a time. The electret core and electron shell would then be rotating together, with the electron shell dominant in determining the direction of the earth's magnetic field. And that is where we are now, with the earth's magnetic field pointing south. (That last point may be confusing for readers who aren't aware that, if the earth is considered a magnet, the *south* pole of that magnet is the one nearer the earth's North Pole, and vice-versa. The north pole of a compass needle is called that because it is "north-seeking.")

I've known all of this for some time and was pleased to read that scientists at Columbia University's Lamont-Doherty Earth Observatory found that the inner core is rotating faster than the rest of the earth. They had no idea what might cause this phenomenon, but I was glad to learn of it since it agrees with what my theory predicts.

One final detail: It is well-known that the angle of the earth's magnetic field is tipped with respect to its rotational axis. The fact that the solid core is not attached to the rest of the earth implies that normal bombardment of the planet by objects from space or other locations within the solar system would not necessarily perturb the core's orientation and motion to the same extent. Over the years, the earth has been tilted vis-à-vis the plane of the ecliptic, while the electret core has not entirely followed suit.

Gravity Was Greater When the Earth Was Smaller

It helps you understand why dinosaurs developed the way they did if you know that the gravity in their day was much greater than today's. These huge animals had massive leg bones to carry their tremendous weight. Certainly the thigh bones of an elephant have to be thick and heavy to carry its weight, but when compared to the thigh bones of a dinosaur of the same size, they are much lighter and less dense. The bones in the upper area of a dinosaur's skeleton, in contrast, are very light. They are also full of holes, which makes them both light and strong, like an airplane wing, whereas the lower vertical bones are massive enough to support tremendous weight. In many of our fine natural history museums, you can compare the bones of modern animals with those of the animals of the Mesozoic.

My favorite example is the dinosaur's spinous process, the bony projection off the back of each vertebra. These projections are exceedingly high columns of bone in the dinosaur, suggesting that the muscles attached to them were very large, which they would need to be to permit the animal to lift a tail made extremely heavy by gravity, and swing it back and forth.

Trees have to be thick at their base to support the weight of the branches and boughs. As it turns out, early trees were shaped with a wider base tapering upward and so were much more substantial at the base than modern trees.

All of these differences are a function of the fact that objects were attracted to the earth much more forcefully when the earth was smaller. The early earth had the same mass, or very nearly the same mass, as it does today. But because it was so much smaller than today, objects on the surface were much closer to the earth's center of gravity. Some expansionists reject the idea that the earth could have had the same mass then as now, but as I explained

Other Mysteries Explained by the Earth's Expansion

earlier, I believe the earth did have nearly its current mass from the beginning, because it was originally a small, dense electret.

Some of the first animals to emerge from the primordial seas had bodies very low to the ground. This orientation was adaptive to the higher gravity present at the earth's surface at the time.

The atmosphere, too, would have been affected by higher gravity. What air there was had to have been denser than it is now, because of the greater force of gravity. The pterodactyl was able to fly by virtue of both its hollow bones and the heavy, dense air.

Gravity may well have a very negative effect on the earth in the future. As I have been saying, the inner core is a solid electret held to its present great density by electrons pressing in on its outer surface. It would not be so dense if there were less pressure. As time passes, the charge within the electret will gradually decrease as atoms escape its surface, thereby reducing downward pressure on it. The present density of the solid core is greater than that of the liquid metal surrounding it, but the elements of which it is composed are much the same as those of the earth's mantle and crust, albeit with many of their electrons missing. There will come a time when the electret's density will be less than that of the metallic outer core. This will constitute only an incremental change for the solid inner core, but it will drastically affect life on the surface of the earth. When the inner core becomes less dense than the outer, it may float off center and thus throw off the balance of the entire planet. With the planet's motions out of kilter, the oceans could not be expected to stay within their shores, and they would flood the earth.

We know that the solid inner core is at present only slightly denser than the liquid outer core, so we could be close to the tipping point. There is really no action to be taken, so it is probably not worth worrying about. It will happen or not, and there is nothing we can do about it.

Chapter 7   Ionite

There is no such thing as ionite, but I didn't know that at the beginning. "Ionite" was the name I came up with for what today I would call an "electret." I invented it within weeks of having realized that the earth expanded, and the fact of that expansion was the only thing I was confident of at the time. I was trying to explain to myself how the matter that makes up the earth could have occupied so much less space than it does now. I thought there had to be a material that was naturally denser than normal matter. My approach to explaining how the earth had expanded was rooted in my being vain enough to assume that I was the only person ever to realize that it had. It turned out I was not, but everyone who previously concluded that the earth had expanded encountered enormous problems when they tried to explain it. If you're in the business of science, you learn early on that, *of course*, the expansion of the earth is impossible. For better or worse, at the time I was just too stubborn to accept that.

A positively charged ion is an atom that has fewer electrons than an atom of the same element that has no charge. If the charge on the ion is great enough—that is, if enough electrons are missing—it is also smaller than an uncharged atom. Even early on, I thought that perhaps our Earth was dense because the matter was somehow ionic. I invented the term "ionite" to describe this hyperdense material. (I was surprised and a little offended when a friend suggested I had chosen the term because it was like my name. We were speaking at the time, and he thought I had said "IoKnight." No, it was always "ionite.") I can't explain how I came up with such a wild notion. At the time, I was certainly desperate to explain expansion. I thought that perhaps the earth was once made up of ions and, by virtue of all the missing electrons, simply did not take up as much space as it does today. That was a crazy idea, since a collection of positively charged ions would simply repel each other, yet this partial insight set the stage for understanding. When I learned of electrets, I realized that there was indeed a form of matter in which a collection of atoms carried a persistent positive charge, but the electrons were not missing. They collected in a halo outside the charged material and pushed inward. I was thrilled to learn that this form of my "ionite" was stable as a solid.

The story behind that realization started years ago, when my wife and I and a guest were all sitting at the dinner table. We had invited a good friend, Peg Repass, who was active at our church and a most interesting person. At some point during the conversation Peg said, "David, tell me how you are coming along with your book"—a good question to ask if one wanted to get me started on a monologue. I had gotten to be a bore on the subject, because I was so enthusiastic about it and didn't know when to stop. At the time, however, I was facing a difficulty. I answered, "Well, actually, I'm stuck with an impossible math problem." Her immediate reply was, "My son John could help you with that. He's very good at mathematics. I'll send him over." John had just graduated from Yale, and he was indeed extraordinarily good at math. John was, and is, very bright.

Despite his gifts, John was lazy in a practical way. One time, he told my wife that he'd chosen not to be an English major at Yale, because he would have had to write too many papers. He'd chosen microbiology instead, since it would require less work. What would have been daunting for most people came easily to John.

Some weeks after our conversation with Peg, John came to our door and said, "Mother sent me over to see you. You have some sort of math problem." I confirmed this and told him what the problem was. By that point, I knew that something like an electret might be able to form, but for all I knew it would simply explode because of the repulsive force of the excess protons it contained. I didn't yet understand how the electret structure could persist, and I was still calling my hypothetical material "ionite." All I was sure of was that it must not explode, because it had to exist as a stable substance if my ideas were correct. John said he didn't know anything about the topics I was dealing with and couldn't even begin to work on my mathematical problem. Well, that was that, I thought. We continued to talk, however, and at some point John said he was interested in learning more about my ideas. He asked if I had anything he could read. Yes, I said, I had been working on the topic for some time and had written 20 or so unpublished papers. I went to my file and handed him my extra copies, about an inch and a half thick when piled together. They represented all my work up to that time. Some pieces were 20 pages long, some only a page. John offered to read a paper each week and come over to talk about what he had read. He did come back, over a period of about six months, and we did talk. It was wonderful for me to be able to convey my thoughts to someone who could understand me and knew what I was writing about.

One evening, after this had been going on for some time, John came to the door. He said with excitement, "Mr. Knight, I think you are right." I will never forget those words. All I could say in reply was, "I thought you thought I was right all along." John said, "Yes, I thought you were right, but now I think you are right." Evidently

Ionite

there is quite a lot of difference between right and right. I asked him to come in and said, "Do you remember why you came to visit me in the first place?" He said, "Yes, you had some sort of math problem." I then showed him all I knew from my research, and we talked about the various forces that must be at play within ionite (electrets). We spent the evening talking, and John told me he would work on the math and return the following week.

When I answered the door a week later, John handed me one of those blue booklets used for writing final exams. In it were 11 pages of math, which at the end showed that the ratio of outward force within the core of an electret to the inward force of the electron halo was three to four. That is to say, reversing the picture, that the electrons surrounding the electret must press inward with enough force to contain the protons inside the electret. John had confirmed that my particles were stable. I could have cried. "Happiness" doesn't come close to describing my feeling at that moment. I told John I would then put together a paper showing his math in a way that scientists might find more concise and helpful. Finally, I had quantitative confirmation that my particles could exist as stable objects in space. Even though the idea itself was simple, it would be new to most physicists, and now I had the means to present the idea in the language of physics, mathematics.

As I reviewed John's math, I found that many of the forces at work within the electret canceled one another out, and it became possible to represent the final result, the net ratio of inward to outward force, as a simple differential equation. All along, I had been making the problem more complicated than it was, because I didn't know how to go about simplifying it.

When I wrote the paper and submitted it to a scientific journal for publication, it was rejected, so I rewrote it and submitted it again. Again it was rejected, and I came to the realization that my discoveries were not going to be accepted, because they were not the way things were *supposed* to happen. If I was right, then conclusions physicists had believed for decades would have to be wrong. I remember someone telling me, "You have a tiger by the tail." I accept that reality for what it is, and I offer my ideas in this book, along with John's invaluable work (see "Appendix: The Math"), so that they may someday be reevaluated on their merits.

Chapter 8   Challenging the Scientific Consensus

The Great Scientists

They have all died. All the scientists whose work informed my thinking about the earth and the cosmos have died, mostly of old age. Like many of them, I was born in the early 20th century, which is one of the reasons I believe I need to publish my ideas now. Some time ago, I read a book that was given to me for Christmas. In it, the Big Bang is listed as one of the errors made by great scientists of the past. Much of the book talks about the persistence of Sir Fred Hoyle in championing his steady-state view of the universe. I know now that he was right in rejecting the Big Bang, and just as importantly, I know *why* he was right. Unfortunately, Hoyle never considered an alternative cause for the red shift and never knew that light loses energy on its way to us as it passes through the electron halos of dust-grain electrets. Still, he was a great man of science who established many facts in astrophysics that we all accept as true, such as his explanation of how fusion in the cores of stars creates all the elements in the periodic table up to iron. He also did not know that the energy lost by light is not wasted, in a sense, because it propels matter out of galaxies to eventually form new ones, and this in turn implies that the universe lasts forever, continually creating new galaxies, star systems, and planets. To me this view seems simplest and best, and what I have tried to do is fill in the blanks of Hoyle's correct belief in a steady-state universe. This still leaves a problem that was there all along: where did all the energy in the universe come from in the first place, since we now know that all mass is energy? I do not have an answer to that question.

Science has both a strength and a weakness, in that a standard of truth is built into it. An idea must by proven many times and in many different ways before it is generally accepted, but once accepted it takes on a life of its own. There seems to be no way to change the minds of scientists once they are convinced of a theoretical scheme, even if it is wrong. This is why Thomas Kuhn's book *The Structure of Scientific Revolutions* came as such a relief to me: it pointed out how scientific paradigms can finally change. Before Louis Pasteur made a case for germ theory, disease was understood to be caused by humours, spirits, or "bad air." Before Antoine-Laurent Lavoisier demonstrated the role of oxygen in combustion, scientists believed that combustible materials contained a fire-like substance

Challenging the Scientific Consensus

called "phlogiston." There were several ideas of this sort that seemed unshakeable yet were changed, primarily through the persistence of one person. Today, we all know that the earth is very old and that it is changing continually over time. James Hutton, the "father of modern geology," was instrumental in bringing about that change in thinking. It took a scientist of strong conviction to overcome the resistance of the earlier paradigm. Even Hutton could not do it alone, but fortunately he found the support he needed.

Today, the geological paradigm in need of revision is the truism that the Earth has always been about the same size. The astronomical one is the Big Bang. With any luck, future scientists will take up my banner and shift the standard thinking in these areas.

Rocking the Boat

It may seem at times that I'm being too hard on science. Yes, I've arrived at ideas our scientists have missed, but I have great admiration for the discipline and do not want to be overly critical. I believe the reason anyone is usually justified in believing what scientists say is that good scientists—the vast majority—are forthright and honest in what they do. They offer their evidence and interpretations of it in good faith. Let us take an example. I know, for instance, that there are electrons. They were discovered by science and are considered real, even though no one has ever seen one. There is research showing that they exist, but our belief in electrons depends on our confidence in those who have done experiments that support their existence. I am using the example of electrons, because they are so uncontroversial. We see evidence of them every day. Many aspects of our everyday life depend on the behavior of electrons, so we accept them as a matter of course. There may be a small number of dishonest scientists who fake data or intentionally put forth false ideas. Fortunately, the most fervent critics of this group of people are the other scientists themselves!

Where do I fall in all this, and what category am I in? I am not a professional scientist, but I believe I am honestly presenting evidence I consider to be factual and interpreting it in a way that I think is correct. In this spirit, I've written about a complete, systematic view of the formation of our universe and our earth. It will be up to you to judge whether or not these ideas have merit.

The other side of the coin about belief is that career scientists know so much and have such extensive backgrounds that it seems presumptuous to pass judgment on them. Anyone with a general-science education knows, or thinks he or she knows, what happened to form our planet. As a result, popular opinion is now my worst enemy. People generally defer to scientific authority for good reason, yet here I am, saying that the authorities have made a mistake. Given that, one might wonder why I undertook writing this book.

I've put together a list of the many areas in which I don't agree with accepted views, which I present in its entirety later in this chapter. At present, I don't think I'm wrong about any of those items, because the whole physical process I describe, although quite different from the standard theories, does make sense. There have been many attempts in the past to lay out a complete system for the cosmos, at every scale, and so far all have failed because of some error. Wherever I have encountered errors in my thinking, I have tried to correct them.

I know there is no such thing as perfection, and I am willing to be proved wrong, but of course I hope it turns out that I'm correct. I believe the most dramatic difference between my view and scientific orthodoxy is the red shift and its cause. If I'm correct on this score, then it follows that the universe goes on forever, that there was no Big Bang, and that there are quasars relatively nearby—all conclusions that would necessarily change our view of the universe. The red shift is considered an infallible tool for measuring the distance of interstellar light sources from our location, but this assumption falls apart if the cause of the shift turns out to be not the movement of the light source but the weakening of the light itself through interactions with halo electrons. I can only hope scientists will explore this possibility in the future.

I also believe we've neglected to take into account the large-scale effects of Van der Waals forces, probably because they have very limited applications here on earth.

My view is that Van der Waals forces, along with gravity, rule the universe in very large contexts. It may be that more will be discovered about them in the future, and the unknowns of astronomy may well be solved with a proper understanding of how Van der Waals forces operate at a distance. I am thinking, for instance, of the way the arms of galaxies are held together. The present enigmas of dark matter and dark energy, too, may simply reflect that we have not taken Van der Waals forces into account.

In an area closer to home, the theory of subduction is the result of another erroneous assumption, as I see it: that the earth has not expanded. No explanation was given for the separation at the mid-ocean ridge when it was found to have pulled apart to form new ocean floor. There *was* new ocean floor, and we knew it had to be going somewhere, but the background belief that we lived on an earth that had always been about the same size prevented us from considering alternative explanations. Of course the reality, I believe, is that at the point where subduction is thought to have taken place, in which the ocean floor is supposed to have been pushed down underneath the continents, there was instead *eduction,* or movement of the mantle in the opposite direction as a result of expansion.

I've thought a lot about how good scientists can be utterly wrong yet convince themselves they are right. If I am right about subduction theory, how could so many scientists have gotten it wrong? The peace-loving side of human nature comes into play. As I've noted before, a seismographic reading cannot tell us in which direction the rock is moving. But if one thinks one already knows what that direction must be, then the interpretation of the data is preordained—and the theory of subduction was already in place when conclusions about these motions were drawn. Anyone suspecting movement in the "wrong" direction would be upsetting a "proven" theory and calling into question the interpretation of all the earlier evidence that went into formulating it. Because most members of a community, including a scientific one, wish to avoid "rocking the boat," accepting dogma becomes the easier choice, and an erroneous theory is only strengthened in each scientist's mind.

After I had decided the earth had expanded and there was no such thing as subduction, an earthquake took place in Alaska. The island of Montague had risen several meters, and an area formerly submerged off its shoreline had become dry land. I saw the event as confirmation of my ideas. But a respected and capable geologist wrote a paper purporting to explain how the earthquake could have caused the observed conditions without contradicting the theory of subduction. He even found a way of explaining the fact that the surface of the mainland, inland from where the earthquake happened, had fallen to a lower level, even as the nearby seafloor rose. From my point of view, if the earth is expanding, then when an earthquake occurs where the seafloor is being pulled out from under the continent, the seafloor rises (in this case, lifting Montague Island), while the continent falls to a lower level. But whereas I saw the quake as evidence of eduction, this geologist managed to write a well-received paper that was consistent with the standard theory.

When plate tectonics theory was first suggested, a Russian geologist disagreed. He asked why the trenches had not long ago filled with detritus if the ocean floor was going down into the trench. Clearly, he was right, but his argument was rejected. There were many attempts to oppose subduction theory, but they couldn't slow down its general adoption. A paper published in *Nature*, for instance, showed that the Pacific was getting larger, but it got nowhere in the geological community. Current plate tectonics has the Pacific getting smaller. Tectonics suggests that the Mediterranean, too, should be narrowing, but in my view it is getting a bit larger as the earth expands. It still amazes me that the present theory remains in place, in spite of all the evidence against it.

My view is that, for two thirds of its existence, the crack in the mantle lay under a continuous crust, and that crack, now filled in and plastered over with rock that rose through it, exists today as the mid-ocean ridge. If I am correct, then the geosynclines of the world created most of our great mountain systems while the mantle crack was still beneath the crust. I know that many who have spent time thinking and writing about the inside of the earth do not share my view. My hope is that science—and you, the reader—will embrace the many ideas presented in this book and build on them, because there is much yet to be done.

Over the years I've spent developing my ideas, many phenomena have been observed that, I believe, had to have been the result of the earth's expansion. One is that the North Pole shifts when an earthquake happens. Because earthquakes of varying intensity are taking place all the time, there is a small change in the position of the Pole each day.

While I was attending a summer program at the University of Chicago during my undergraduate years, I learned from a fellow student that this change in the position of the geographic poles was correlated with earthquakes. He had worked for the U.S. Bureau of Standards (renamed in 1988 as the National Institute of Standards and Technology, or NIST) some years before, where his daily job was to determine mathematically, using telescope, paper, and pencil, the exact location at which the line of the earth's axis of rotation met its surface—i.e., the geographic pole (in this case, the North Pole). Each day, he would make his calculations twice for accuracy. One day, on his second try, he found that the pole had moved almost two feet. He rechecked his work and could find no error. At a loss to explain the discrepancy, he reported both results. It turned out there had been a significant earthquake during the interval between his calculations.

Much later, after I concluded that the earth was expanding, I realized what was happening. The earth does not expand evenly. Rocks within the earth adjust to expansion in fits and starts, and those adjustments manifest themselves as earthquakes. As rocks shift position with each earthquake, their movement slightly changes the earth's center of mass, which in turn shifts its axis of rotation.

Ideas of Mine That Run Counter to Accepted Theory

1. There is only one universe.
2. The stars and universe are not the result of string theory.
3. There are no black holes.
4. Dark matter does not exist. There are other forces at work in deep space that we have not yet considered.
5. The universe has no limit and goes on forever in all directions.
6. There is no limit to time. Time extends forever into the past and goes on forever into the future.
7. There was no Big Bang. In this, Einstein's first thought was the right one.
8. Nothing can go faster than the speed of light, making Einstein right again. (This *is* generally accepted, but there is today a contrary school of thought.)
9. There are no voids in space.
10. The "red shift," or the change in the wavelength of light from distant stars toward the red end of the

spectrum, is not caused by the stars' rapid motion away from us. Rather, light loses energy as it travels through space, when it encounters electron halos around charged cosmic dust.

11. Both Einstein and Hubble were correct about different aspects of the red shift. It was Hubble's use of Einstein's findings that was the error.
12. Interstellar background radiation is very old light that has been lengthened as it has passed through electron halos around charged dust particles.
13. Quasars are not located at astronomical distances away from us but rather are closer by.
14. The cosmological constant, which Einstein added to his general theory of relativity in order to allow for a static universe, is not necessary after all.
15. In time, all stars explode to become supernovas.
16. All stars were once surrounded by rings early in their formation.
17. A planet's rings contain charged dust surrounded by electron halos.
18. Van der Waals forces act between charged electret dust as they act between neutral gas atoms.
19. Within clouds of charged space dust, Van der Waals forces function at a distance. Van der Waals forces, together with gravity, produced our solar system.
20. Many comets are small clouds of dust-particle electrets that happen to fall into our solar system.

21. A meteor that is visible as a "shooting star" is a charged group of electrets—a comet or a piece of a comet—falling into our atmosphere. The light of a shooting star is a meteor's electrets being neutralized as the meteor heats up in the atmosphere. (Meteorites, generally, are pieces of other solid bodies such as planets, moons, or asteroids.)
22. The earth has expanded, and continues to get larger because of the decay of its inner electret core. (Geologists were provided with evidence of the earth's expansion sixty years ago but did not believe it was possible.)
23. The inner core is all that is left of the original electret earth.
24. The Pacific Ocean continues to increase in size.
25. The Mediterranean Sea is widening, not being pushed together.
26. India did not move rapidly across the ocean and push up the Himalayan Mountains.
27. The ocean *floor* is the top of the mantle covered by sediment, whereas an ocean *rise* is composed of crust-type rock, and is thus really a thin, submerged continent.
28. There is a layer of electrons between the earth's inner and outer core.
29. The earth's inner core is free to turn within the electron shell at a different rate than the rest of the earth.

30. The earth's magnetic field is not due to churning within the mantle; it is due to the rotation of its electret core and the shell of electrons surrounding the core.
31. The earth's magnetic field can be, and has been, reversed by large meteor strikes.
32. The crack in the mantle due to the earth's increase in size caused the crust to stretch and created the various types of rock that characterize the geological periods.
33. When the Americas first pulled away from Asia, what is now the Shatskiy Rise was made up of debris that had washed off Asia and filled the resulting depression in the crust. The rise later pulled away from the continent as the sea floor also moved away because of the earth's expansion.
34. In the late Permian Period, a giant glacier covering the southern part of the earth's continuous crust ended up holding nearly all the water that existed on earth at the time. (There were as yet no oceans.) This led to the "Great Dying" (the Permian-Triassic extinction event) and the end of the Paleozoic Era. In other words, the fish didn't leave the water, the water left the fish.
35. The rate of the earth's expansion increased to a maximum in the Mesozoic, then decreased to its present rate.

36. The position of the mantle is relatively unchanging, so its crack (now the mid-ocean ridge) remains nearly where it formed.

You no doubt realize by now that there is much more I have left unsaid. I've learned a great deal as I've gone along, keeping track of the many details in papers and an earlier, longer version of this book. This time, I've kept the story short and, I hope, readable. I've been told that some of my earlier writings make for difficult reading. My writing, like my ideas, has changed over the years.

Chapter 9   Final Thoughts

After I realized that I was unlikely to have a paper describing my theory accepted for publication by a scientific journal, a wise friend admonished me, "It is your thing, and you do not have to involve anyone else for now." What he said rang true at the time. But I'm now an old man, and I recently realized that if I was going to get these ideas of mine out into the world, it had to happen soon.

Yes, the conclusions I've presented here all started with the realization that the earth had expanded, and I soon found out that the rest of the world thought otherwise. Exactly when I realized what must have taken place and how the many pieces fit together, I don't know, but it wasn't long ago that the whole picture came together.

Often, I've advanced my thinking after being asked a question I hadn't considered. I couldn't always provide an answer at the time the question was asked, but after working at it for a while I'd be able to add another piece to the larger puzzle.

I've often said I did most of my best work in the morning. There were times I would write, and words would just appear on the page. I would come up with my best ideas in bed at four a.m., then write them out more fully and precisely later on. It seems that somehow my subconscious had worked out the problem while I was asleep, without my conscious mind at first knowing what my subconscious was thinking.

I don't know everything that has happened by a long shot, but I do know now that the earth was formed as a charged mass that was held to its small size by electrons on the outside. I am confident it happened this way because the story fits reality perfectly. Every time I've encountered a truly new piece of information, it has been a simple thing for me to account for the new information within my scenario. The phenomenon of the earth's magnetic field, which I discussed in Chapter 6, is a perfect example of this.

I've wondered about the earth's magnetic field for many years, and of course I haven't been alone. Long before I discovered the expanding earth, I remember going to visit a friend who had an Encyclopedia Britannica, which allowed me to check an idea I came up with on this topic. That idea turned out to be completely wrong, but at least I had started thinking about issues of this sort. As it turned out, I wouldn't understand what caused the magnetic field until I knew that the central core of the earth was an electret. And I wouldn't know what caused the historical reversals of the field until I had realized there was a layer of electrons above the electret core. I still do not know just how slight the friction is within that electron layer; my basic conclusion is just that there must be very little. That's because electrons have so little mass. I haven't done any experiments to get a precise measurement. I don't even know how one would set up such an experiment, since it would require creating a layer made up only of electrons between two surfaces, and no such thing occurs naturally at the surface of the earth.

When I first conceived of this layer of electrons around the core, I couldn't believe it could exist. It took me a long time to accept that such a condition could really be possible. This is one of those notions I would lie in bed thinking about at four in the morning, before I got up to go to work. My sequence of thought was, first, that in order to exert pressure on the surface of the electret, the electron layer would have to be pulled in with tremendous force. The electrons would be unable to force their way into the electret, and since each electron would have to press against its neighbor with equally great force, the electrons in this layer would end up being very close together. Next, I considered that if they pressed down on the electret and also against one another, they must press *upward* as well. The conclusion from this was that their density would have to diminish quickly with distance away from the surface of the electret. Then, with the realization that the metal of the liquid core above could not be as dense as the layer of electrons right next to the electret core, I then believed the electron layer could exist.

By the way, we know that the electret must be denser at the core's surface than the concentration of electrons farther away. The surface of the core must be super-dense, in order to prevent the invasion of electrons. This is further evidence that the inner core must be denser than the liquid core.

I have tried to explain the existence of our solar system and the formation of the nascent earth. The philosopher Kant said that the solar system started as a cloud of dust. I think he was right. My advantage has been that I found out where the dust came from, or at least about the halos surrounding each grain of dust in the original cloud. I didn't start with the cloud but was able to infer that there *would* be a cloud, given that there were charged dust particles. I had a further advantage in that I knew the whole process was going to result in a compact planet much smaller than the earth now but with the same mass.

Only gradually did I come to understand that the solar system started out as a disk that broke apart into rings that coalesced, as I have described, to form the planets. Some were solid and compact, like the earth, while in others the charged particles did not become solid bodies but rather formed the so-called "gas giants," Jupiter, Saturn,

Final Thoughts

Uranus, and Neptune. It was quite recently that I figured out how the rings changed into the planets. My description of the way the sun gradually developed and the decay of the rings around it was just the result of one idea following the next. It was not all thought out first, then written down. I only followed the trail of how events would logically unfold, keeping in mind a few facts from physics and other fields.

Everything you've been reading about the earth and the cosmos has been the product of this kind of step-by-step reasoning. My version of the history of the universe and our planet is not nearly as complicated as what is commonly thought to have taken place. All that I've accomplished here has been built on the achievements of great men and women of science. As Will Rogers put it, "I only know what I read in the papers," the "papers" in my case being the hard work of those great scientists.

My contribution has been to sort out some of the truth from the errors. I decided that the earth had expanded. I was not the first to arrive at this conclusion—Carey made the same determination a decade earlier—but I did think of it myself and am proud of that fact. It then took me many years to realize when the crucial mistake had been made: when Hubble adopted Einstein's explanation of the red shift, an assumption that led to the Big Bang and a universe of limited size.

Many years ago, a young teacher in my town heard through our church that I was working on the expanding earth. He gave me a textbook, which I still have, *Study of the Earth: Readings in Geological Science*, by John White. It covered the knowledge available at the time and was a good start. Early on, when I realized that very few people believed the earth had expanded, it seemed pointless to buck all of established science. It looked as though I would have to fight for my ideas, and that is not in my nature.

The real reasons for my writing this book now are that I'm not going to live forever and that my overall view finally came together once I came to understand Hubble's crucial error. Fritz Zwicky correctly asserted that that reason for the red shift was that light loses energy on its way to us, but his suggestion was rejected because he thought the energy was lost to dust particles, a process for which no one had an explanation.

In geology, similarly, the scientific community has ignored the few colleagues, including S. Warren Carey, who have tried to make the case that the earth expanded. Of course, those earlier expansionists did not know precisely *how* the earth expanded, but they were on the right track. I am not bitter about this, but I am disappointed in our geophysicists for rejecting alternative ideas out of hand. All I can do now is write about what actually goes on, as I have tried to do.

I've always loved the sciences and still do. Yes, scientists have been wrong, but eventually they tend to ac-

Final Thoughts 115

cept the way things are, when they have seen all the evidence. I hope our scientific community will someday come around to my way of thinking. As I said, I am not a fighter, and the time when I might have enjoyed fame and glory is long past. If my ideas were accepted now, it would only spoil my very comfortable life. I am too old to get any thrill out of acclaim. What you read here is just what I believe did happen.

My work on this has taken most of my adult life. It has been both a hobby and an introduction to the way science works. I can understand why our professional scientists did not find support for the earth's expansion. In order to uncover the causes behind it, they would have had to upend their own convictions. They would have had to discover how the sun and its system of planets and other bodies actually formed. They would have had to consider how Van der Waals forces might operate at a distance. They would have had to find how dust particles can become charged and develop halos of electrons. They would have had to get used to the idea of electrons pushing solid matter into a small volume. Scientists are not used to thinking of electrons behaving this way. They also would find it difficult to imagine solid rock being compressed to a fraction of its normal size.

I found out how and why quasars are actually near us, not out at astronomical distances, yet I could not convince Halton Arp, the man who had determined this mathematically, why he was right. Hoyle knew that there was no Big Bang and spent many years trying, unsuccessfully, to prove it.

I was sure, too, that many comets are clouds of electret particles of different sizes and at different stages of development. I realized that it was a small comet that fell into our atmosphere in Tunguska, Siberia many years ago, causing an electrical explosion above the earth's surface in which halo electrons rejoined charged material from within the exploding comet's electret to neutralize it, leaving behind particles of ordinary matter but no impact crater and no meteorite.

The causes of the earth's magnetic field and its reversal were obvious, once I knew that the inner core was a highly charged, compressed electret.

Good men and women have worked on the theory of the inside of the earth, but they could not know the real reason the earth expanded or how the crack in the mantle was the primary cause of the formation of continents and of the various geological periods and eras at the surface.

As I have said before, this entire project started with my realization that the earth must have expanded. The problem is that, since that turning point, I have discovered a number facts that no one else knew and that have not been accepted by the scientific community. I am afraid scientists will not investigate my findings and so will postpone their own eventual discovery of what I have learned. Perhaps it can't be otherwise. Too many of their beliefs

Final Thoughts

would have to change—*they* would have to change—before they could accept the information I offer here.

Each new fact I discover is simply added to my fund of knowledge, and it no longer bothers me that others do not see things as I do. Yes, I realize some of these facts will one day be considered "breakthroughs," but their lack of popularity now does not change the way things are. I am thinking particularly of the implications of a cloud of electrons surrounding an absolute nonconductor such as the inner core of the earth. It took me some time to get used to the idea of the tremendous force that electrons can exert on matter.

I broke the rules. I suggested that the limit to the amount of charge that can be carried by a piece of matter here on earth does not apply in deep space. Out there, charges embedded in a grain of dust have no other object to jump to. An object in space could therefore maintain a tremendous charge, however unlikely that may seem. We have to think differently. It is my hope that you, the reader, will come to accept my view of the universe as the way things really are.

The laws of physics still apply under my view, with some added details. Van der Waals forces were discovered many years ago, but we have not thought through the effect they might have in space, in the absence of competing forces. Space dust particles become surrounded by electrons through a process I think takes place all the time. I suspect Van der Waals forces act between these charged particles because of their electron halos. We never think of that possibility, because we examine those forces only on earth.

In geophysics, one of the current assumptions that I am sure will become a problem is the idea of subduction. I have had no problem rejecting that theory, but it will sooner or later become a real problem for the seismologists who now accept it.

I was fortunate to have discovered that the earth expanded at about the same time it was discovered that the ocean floor was moving away from the mid-ocean ridge. This was where Harry Hess came into the picture. I've mentioned that it was his idea of seafloor spreading that started me thinking. I was his opposition, but of course he did not know I even existed. I lived through the entire development of the theory of plate tectonics. I subscribed to the New York Times, because the Tuesday science section always used to have something to say about geology. Unfortunately, nowadays the Times and everyone else seem to think the important problems of geology have been solved. The Times no longer writes about what goes on inside the earth. It is not that the people there are completely uninterested, but the earth today is not news, and news is their business. It is my hope that I will get them going again if they read this little book. (As I write this, tomorrow is Tuesday, by the way, and I still subscribe.)

It has become common knowledge that the universe started with the Big Bang, but here I am, saying there was

Final Thoughts

no Big Bang. I am saying that the universe has been going on forever, and I give a description of how its process of self-creation unfolds, again and again. No, I don't know how that happened or where all the energy in the universe came from, so in a sense we really are back at the beginning. Einstein at first thought the universe had no beginning, but after meeting Hubble he was convinced, unfortunately, that the universe was expanding.

I still admire the scientific spirit, because scientists can change their minds when they are convinced by new ideas. Sometimes it takes a lot of convincing. I remember hearing as a child that scientists thought the world was two billion years old. A year later, they would say they had misread their data; it was really four billion years old. Now we are quite sure it is some 4.6 billion years old. And this is before they have read my book. We will have to see how much they add to their estimate to take into account the accumulation of the earth's matter from charged dust into its original, highly compacted form.

Perhaps this is not the time or place for me to wax philosophical, but if you truly believe I am correct, then you must be prepared to be called some sort of a nut. Don't be discouraged; you're not a nut. The earth expanded from a small, dense sphere, and there was no Big Bang.

At this point, I don't know what else to say. What I have done is work out how the earth formed, along with a number of additional conclusions that flow from that explanation, and there is little more that I really know. In coming to this point, I spent a great deal of time imagining myself speaking about what I knew to be true. It was a way of working, although I often did it as I was falling sleep. I would try to arrange the things I knew to be true and present them to an invisible audience. Even though this method has proven fruitful, certain ultimate answers still elude me. As I said, I don't know where all of the energy came from that formed the universe. (I accept without question that energy can become matter and vice versa.) Given these uncertainties, I'm not sure how I should have titled this little discussion. It could be thought of as an introduction to further exploration, but I understand there is probably a limit to what mankind can know. This may be why belief in a supernatural being is so prevalent and is no doubt in our genes.

Based on what I know now, I can say at least that the universe is very large, yet we human beings do exist in it, because we are the ones capable of pondering both it and ourselves. Descartes seems to have been correct in reasoning, "I think, therefore I am."

Appendix   The Math

Below, I lay out the math that was first worked out by John Repass many years ago. As I explained earlier, I have revised and simplified what he first wrote. At the time, I wasn't sure my hypothetical "ionite" particles would be stable. John was able to show mathematically that they were, which allowed me to go on from there. I discovered later that these particles were electrets. I've tried to write the math so that anyone can read it. If you're able to manipulate figures you can prove for yourself that it works out.

What I have done is to first define each term and then to write out the computation, with what I hope is a clear description of each step. For anyone with a mathematical bent this should not be too difficult.

Notation used throughout

R	=	Radius of a spherical electret.
Q_c	=	Total positive unbalanced charge within the electret.
Y	=	Charge density of homogeneously distributed charge within the electret.
Q_e	=	Total negative charge, equal in magnitude to Q_c, which is attracted to the non-conducting outside surface of the electret.
P_e	=	Inward pressure due to the electrons pressing in on the outside surface of the electret.
A	=	Area of the outside surface of an electret of radius R.
P_c	=	Outward pressure created by the repulsion of confined positive charges within the electret.
P_n	=	Excess of P_e greater than P_c, so that $P_e - P_c = P_n$ gives the hydrostatic pressure within the electret.
r	=	Variable radius of a hypothetical spherical shell within the spherical electret concentric with the sphere.
dr	=	An infinitesimal distance between a second, outer concentric spherical shell and the shell of radius r just inside it.

Appendix: The Math

$r + dr =$ Radius of an outer concentric spherical shell infinitesimally larger than an inner one of radius r.

$Q_r =$ Charge within the inner shell.

$Q_r\, dr =$ Charge within the space between the first and second shells.

$F_r =$ Outward acting force due to repulsion of positive charges within a sphere of radius r (i.e., the inner shell) acting against the positive charges within the volume dr thick between the inner and outer shells.

$F_c =$ Total outward force due to the repulsion of all the positive charges acting against one another within the electret.

$F_e =$ Total inward force of electrons pushing in on the outside surface of the electret because of attraction toward its center.

$k =$ Electrostatic constant used, roughly 9×10^9 newtons times meters squared per coulomb squared, in the MKS system of measurement units (now the SI system).

In a spherical dielectric of radius R, let there be a homogeneously distributed unbalanced charge Q_c such that the charge density is Y charges per unit volume. It is therefore a positive spherical electret where the total charge Q_c equals Y times the volume, so that

$$Q_c = Y\left(\frac{4}{3}\right)\pi R^3. \tag{1}$$

Free electrons will be attracted to the outside so that their total negative charge Q_e is equal in magnitude to the total unbalanced positive charge Q_c contained within the electret. The charged dielectric is stable if it is found that the pressure inward, P_e, exerted by the surrounding electrons, with total negative charge Q_e is greater than the outward pressure P_c due to the mutual repulsion of individual positive charges which total Q_c. The electret is physically unstable only if P_c exceeds P_e, but it can be shown here that $P_e/P_c > 1$, regardless of the size of the charge (Q_c) or the radius R.

Consider within the original sphere two hypothetical concentric spherical surfaces, an inner sphere of radius r and a second of radius $r + dr$, with its surface an infinitesimal distance dr outside the first. Since dr is small, the volume charge of solid dielectric contained within this interstitial space is equal to the area of the inner shell (the sphere of radius r) multiplied by the charge density, so that

$$Q_r dr = Y 4\pi r^2 dr. \tag{2}$$

Consider another charge Q_r, which is the total of the unbalanced positive charges contained within the sphere of radius r. These charges totaling Q_r act in concert, as though they were all concentrated at the exact center, to

Appendix: The Math

repel outward all of the unbalanced positive charges totaling $Q_r dr$ contained within the volume between the two concentric shells, the inner one of radius r and the outer one of radius $r + dr$. The magnitude of Qr is equal to the product of the volume of the sphere of radius r and the charge density, Y, so that

$$Q_r = Y\left(\frac{4}{3}\right)\pi r^3. \tag{3}$$

From Coulomb's Law, the repulsive force F_r between the central sphere charges, Q_r, and the charges in the shell, dr thick, is

$$F_r = \frac{[k(Q_r)(Q_r dr)]}{r^2}. \tag{4}$$

Substituting, we find

$$F_r = kY^2\left(\frac{16}{3}\right)\pi^2 r^3\, dr, \tag{5}$$

where k is the electrostatic constant.

The charges in the rest of the sphere, between the outer shell at radius $r + dr$ and the surface at radius R, are not found in the force equation. All this remaining volume with its positive charges can be considered a set of charged concentric spherical shells where, from Gauss, each charged surface creates no effect on any charges within its particular volume.

We can therefore write a differential equation for the total outward acting force, F_c, as r goes from zero to R for the volume of the sphere of radius R that ignores those external charges so that

$$F_c = kY^2 \left(\frac{16}{3}\right)\pi^2 \int_0^R r^3 \, dr. \tag{6}$$

Integrating, we find:

$$F_c = \frac{kY^2 4\pi^2 R^2 \times R^2}{3}. \tag{7}$$

This force outward, F_c, divided by A, the area of the square, radius R, is the theoretical outward acting pressure P_c created by the repulsion of all positive charges by each other. Therefore

$$P_c = F_c/A \tag{8a}$$

or

$$P_c = \frac{\left(\frac{4}{3}\right)kY^2\pi^2 R^4}{4\pi R^2}. \tag{8b}$$

This simplifies to

$$P_c = \left(\frac{1}{3}\right)kY^2\pi R^2. \tag{9}$$

P_c is still not enough pressure to overcome the inward pressure P_e that holds the mass intact, even neglecting the dielectric strength of the solid, which might be expected to diminish the effect of the internal outward acting pressure. The physical integrity of a highly charged electret is therefore maintained by the wall of electrons attracted to the outside surface pushing to get in.

Since, as noted above, Q_e has a magnitude equal to Q_c, then again from Coulomb's law ($F_e = kQ_eQ_c/R$) and because $Q_c = Q_e = (4/3)Y/R$ and because $Q_c = Q_e = Y(4/3)\pi R$, then

$$F_e = \frac{k\left[Y\left(\frac{4}{3}\right)\pi R^3\right]^2}{R^2}. \tag{10}$$

This simplifies to

$$F_e = kY^2\left(\frac{16}{9}\right)\pi^2 R^4. \tag{11}$$

As noted, $F_e/Area = Pressure$, so we have

$$P_e = \frac{kY^2(16/9)\pi^2 R^4}{4\pi R^2}, \tag{12}$$

which becomes

$$P_e = 4/9\, kY^2 \pi R^2. \tag{13}$$

As noted earlier, in order that the electret always be stable, P_e/P_c must always be greater than one. The ratio of pressure inward (P_e) to pressure outward (P_c) is

$$P_e/P_c = \frac{(4/9)kY^2\pi R^2}{(1/3)kY^2\pi R^2}. \qquad (14)$$

($kY^2\pi R^2$) cancels out of both P_e and P_c, leaving

$$P_e/P_c = 4/3. \qquad (15)$$

And this value is greater than one. As Archimedes said, "Eureka!"

Thus an electret charged to a high degree remains stable regardless of its charge. There are limitations, of course, since if there were too many electrons missing there would be no matter to keep stable, and to this we must add the requirement that the electrons must not penetrate and thereby neutralize the internal charges. Under certain conditions, therefore, a solid dielectric could exist with a large positive charge distributed more or less homogeneously throughout its mass, so long as there was also a supply of free electrons that are allowed to accumulate at its surface.

The electret might well be unstable if it were charged through some process of ion implantation but not permitted to accumulate the necessary free electrons. Obviously, positive ions do strongly repel each other, and at the point where the physical strength of the dielectric is reached, if not for the electrons outside, one might expect it to burst.

We are left with the fact that if there are free electrons available, and if the material does not spontaneously discharge, then there will be an excess of pressure due to the electrons pressing in on the outside surface of the highly charged, non-conducting electret. If P_n is this excess pressure, then

$$P_n = P_e - P_c, \qquad (16)$$

so that

$$P_n = P_e - P_c = [(4/9 - 1/3) \times (kY^2 \pi R^2)], \qquad (17)$$

which yields

$$P_n = .111(kY^2 \pi R^2). \qquad (18)$$

Note that P_n is applied as a real hydrostatic pressure within the body of the electret and that it increases proportionally to the square of the electret's radius (R) as well as to the square of the charge density.